CBEST Prep Book with Practice Tests Collection for California Educators

CBEST Math, Reading, and Writing Study Guide

CBEST and the California Basic Educational Skills Test are trademarks of the California Commission on Teacher Credentialing, National Evaluation Systems Inc, and Pearson Education, Inc, which are not affiliated with nor endorse these practice tests.

CBEST Prep Book with Practice Tests Collection for California Educators: CBEST Math, Reading, and Writing Study Guide

© COPYRIGHT 2015, 2021. Exam SAM Study Aids & Media dba www.examsam.com

All rights reserved. No part of this publication may be reproduced, stored in a retrieval system, or transmitted, in any form or by any means, electronic, mechanical, photocopying, recording, or otherwise, without the prior written permission of the copyright owner.

ISBN: 978-1-949282-71-9

NOTE: The drawings in this publication are for illustration purposes only. They are not drawn to an exact scale.

For information on bulk discounts, please contact Exam SAM via our webpage.

CBEST and the California Basic Educational Skills Test are trademarks of the California Commission on Teacher Credentialing, National Evaluation Systems Inc, and Pearson Education, Inc, which are not affiliated with nor endorse these practice tests.

TABLE OF CONTENTS

CBEST MATH

Format of the CBEST Mathematics Test	1
Types of Questions on the CBEST Math Test:	
Estimation and Measurement	2
Statistical Principles	2
Computation and Problem Solving	2
Numerical and Graphic Relationships	2
How to Use the Math Study Guide	4
CBEST Practice Math Test 1:	
Interpreting data presented in tables	5
Performing calculations on fractions	6
Performing calculations on percentages	7
Calculating test scores and determining averages	8
Estimating results without doing computations	8
Numerical relationships with "less than" or "greater than"	8
Measuring solid and fluid weights	9
Measuring length and perimeter	9
Performing calculations on whole numbers	10
Solving practical problems	10
Interpreting standardized test scores	11
Working with positive and negative numbers	12
Identifying required facts in problems	12
Basic probability and making predictions	13
Setting up algebraic equations	15
Solving algebraic equations with one unknown variable	16
Identifying and using information in data charts	17
Calculating distance traveled	22
Calculating number of units sold	23
Determining if sufficient information is provided to solve problems	24
Calculating prices per unit	24
Identifying mathematical equivalents	25
Rounding whole numbers and decimals	26
Word problems on logical relationships	27
Calculating discounts	28
Using percentages to calculate a tax increase	30
Ordering fractions from least to greatest	30
Identifying and using information in line graphs and pie charts	32

CBEST Practice Math Test 2	36
CBEST Math Practice Test 2 – Answers	48
CBEST Practice Math Test 3	60
CBEST Math Practice Test 3 – Answers	73
CBEST Practice Math Test 4	82
CBEST Math Practice Test 4 – Answers	95

CBEST READING

Format of the CBEST Reading Test	102
Types of Questions on the CBEST Reading Test	102
Critical Analysis and Evaluation Questions	102
Comprehension and Research Skills Questions	102
How to Use the Reading Study Guide	103
CBEST Practice Reading Test 1 – Questions, Tips, and Explanations:	
Using an index	104
Identifying where specific information is located within a book	104
Understanding the author's purpose	105
Determining the meaning of unknown words	105
Drawing conclusions	105
Summarizing information from a passage	106
Using key transition and linking words	107
Recognizing the main idea	107
Identifying the style of writing in a selection	107
Identifying details from a passage	108
Recognizing the intended audience for the selection	108
Understanding the author's viewpoint	108
Discerning facts from opinions	109
Determining the relevance of ideas to the selection	109
Interpreting information from a table, chart, or graph	110
Determining the correct order of events or steps in a process	111
Arranging ideas logically within a selection	111
Drawing inferences from information in a selection	111
Understanding the author's persuasive technique or strategies	112
Ascertaining the relationship between general and specific ideas in a selection	113
Understanding the organizational scheme of a selection	113
Ascertaining the relationship between meaning and context	114
Identifying the author's assumptions	114
Comparing and contrasting the ideas stated in the question to those within the selection	115
Identifying varying interpretations of a word	118

Finding the title of a selection	118
Using a table of contents	121
CBEST Practice Reading Test 2	127
Answers and Explanations to Practice Test 2	141
CBEST Practice Reading Test 3	146
Answers and Explanations to Practice Test 3	161
CBEST Practice Reading Test 4	165
Answers and Explanations to Practice Test 4	178

CBEST ESSAY WRITING

PART 1 – About the CBEST Essays:

CBEST Essay Format & Question Types	183
The CBEST Expository Essay Task	183
The CBEST Personal Experience Essay Task	184
CBEST Essay Scoring – How Your CBEST Essays Are Marked	184
How to Avoid Common Essay Errors and Raise Your Score	186

PART 2 – Writing the CBEST Expository Essay:

CBEST Expository Essay Structure	187
Creating Effective Thesis Statements	188
Thesis Statement – Exercises and Answers	188
Writing the Introduction	190
Writing the Introduction – Exercises and Answers	191
Organizing the Main Body	193
Elaboration in the Body Paragraphs	194
Elaboration of Supporting Points – Exercises and Answers	195
Writing the Main Body Paragraphs – Exercises and Answers	197
Writing Clear and Concise Topic Sentences	199
Topic Sentences – Exercises and Answers	200
Writing the Conclusion	202
Writing the Conclusion – Exercises and Answers	203
Sample Expository Essay – Model Essay 1	205
Sample Expository Essay – Model Essay 2	206
Analysis of Model Essay 2	207
Sample Expository Essay – Model Essay 3	208
Analysis of Model Essay 3	209

PART 3 – Writing the Personal Experience Essay:

CBEST Personal Experience Essay Structure	210
CBEST Personal Experience Essay Tips	211
Useful Phrases for the Personal Experience Essay	212
Sample Personal Experience Essay – Model Essay 4	213

Analysis of Model Essay 4	214
Sample Personal Experience Essay – Model Essay 5	215
Analysis of Model Essay 5	216

PART 4 – CBEST English Grammar Review:

Avoiding Misplaced Modifiers	217
Negative Inversion	217
Past Participle Phrases	218
Past Perfect Tense	218
Pronoun-Antecedent Agreement	218
Pronoun Usage – Correct Use of *Its* and *It's*	219
Pronoun Usage – Correct Use of *Their*, *There* and *They're*	219
Pronoun Usage – Avoiding "You" and "Your"	219
Pronoun Usage – Demonstrative Pronouns	220
Pronoun Usage – Relative Pronouns	220
Punctuation – Avoiding the Parenthetical	221
Punctuation – Using the Apostrophe for Possessive Forms	221
Punctuation – Using Colons and Semicolons	221
Punctuation – Using Commas with Dates and Locations	221
Punctuation – Using Commas for Items in a Series	222
Punctuation and Independent Clauses	222
Punctuation and Quotations Marks	222
Restrictive and Non-restrictive Modifiers	223
Sentence Fragments	223
Subject-Verb Agreement	223
Grammar and Punctuation Exercises	224
Grammar and Punctuation Exercises – Answers	225

PART 5 – Developing Your Sentences:

Using Phrases, Clauses and Cohesive Devices	226
Sentence Linkers	226
Phrase Linkers	226
Subordinators	227
Cohesive Devices by Category	228
Sentence Development Exercises	231
Sentence Development Exercises – Answers & Explanations	234

PART 6 – Essay Correction Exercises   239

Answers to the Essay Correction Exercises	257

CBEST MATH

Format of the CBEST Mathematics Test

Three types of skills are assessed on the CBEST test of mathematics:

- Skill 1: The first skill group assessed on the exam is estimation, measurement, and basic statistical principles.

  o Estimation problems involve rounding figures up or down and then performing addition, subtraction, multiplication, or division.

  o Estimation problems also involve making estimates of time to complete a task or a journey.

  o Measurement problems cover knowledge of dimensions or length, weight, and capacity.

  o Measurement problems on the CBEST often include diagrams.

  o Statistical problems on the exam are normally related to the calculation of test scores, although there may be other types of statistical questions.

- Skill 2: The second skill group covered on the CBEST math test is computation and problem solving.

  o Computations will involve addition, subtraction, multiplication, and division.

  o Problem solving questions will normally present a practical problem for you to work out, such as calculating the amount of discount on an item on sale.

- Skill 3: The third skill group on the test is numerical and graphic relationships.

  o Numerical relationship questions will ask you to determine whether a given number is less than or greater than other numbers.

  o For graphic relationship questions, you will usually have to interpret data from a table, chart, or graph.

Types of Questions on the CBEST Math Test

Estimation and Measurement

The CBEST exam includes these types of estimation and measurement questions:

- Understanding how to measure temperature, length, weight, and capacity using United States' measurement systems
- Measuring linear distances and perimeter
- Estimating the time required in order to achieve a work-related objective
- Estimating the results of problems involving addition, subtraction, multiplication, and division without doing full computations

Statistical Principles

Statistical questions on the CBEST cover the following skills:

- Performing arithmetic on basic data related to test scores, such as averages, ratios, and percentages
- Interpreting standardized test scores such as stanine scores and percentiles
- Using test scores to understand how a particular student has performed relative to other students
- Understanding basic probability to make predictions based on the data provided

Computation and Problem Solving

Computation and problem-solving questions will cover these skills:

- Adding, subtracting, multiplying, and dividing
- Performing calculations on whole numbers, both positive and negative
- Performing calculations on fractions, decimals, and percentages
- Solving practical problems, such as determining prices per unit
- Solving algebraic equations with one unknown variable
- Determining if enough information is provided in order to solve a problem
- Identifying the facts given in a problem
- Understanding alternative methods for solving problems

Numerical and Graphic Relationships

The exam includes these types of numerical and graphic relationship questions:

- Recognizing relationships in data, such as calculating a percentage increase or decrease
- Ordering fractions from greatest to least or least to greatest
- Determining if a given number is less than or greater than other numbers
- Using less than, greater than, and equal to express mathematical relationships

- Identifying mathematical equivalents, such as $1/5$ equals 20%
- Using rounding to solve problems
- Understanding word problems that contain logical relationships, such as "if-then" sentences or quantifiers like "some" or "no one."
- Identifying data that is missing from a table or graph
- Using data in tables, graphs, or charts to solve problems

How to Use the Math Study Guide

The practice math tests in this study guide contain questions of all of the types that you will see on the real CBEST test.

Practice test 1 in this book is in "tutorial mode."

As you complete practice test 1, you should pay special attention to the tips located in the special boxes.

Although you will not see tips like this on the actual test, these suggestions will help you improve your exam performance.

You should also study the explanations to the answers to practice test 1 especially carefully.

The tips that you will see in the questions and explanations to math practice test 1 will help you obtain strategies to improve your performance on the other practice tests in this book.

Of course, these strategies will also help you do your best on the day of your actual CBEST math test.

CBEST Practice Math Test 1

1. Use the table to answer the question that follows.

Skill Area	Total Possible Points	Points Received
Flexibility	15	7
Strength	30	28
Speed	35	31
Stamina	20	16

   An athlete takes part in a gymnastics competition and receives the points indicated in the table above. What percent of the total possible points did the athlete receive?
   A. 28
   B. 31
   C. 72
   D. 82
   E. 92

   > Question 1 is a numerical and graphic relationship question which involves using data presented in a table format. You have to decide which data to use from the table in order to solve the problem.

Tips and Explanations:

1. The correct answer is D.
   STEP 1: First of all, you should add up the total possible points as shown.
   15 + 30 + 35 + 20 = 100 possible points
   STEP 2: Then add up the points the athlete received.
   7 + 28 + 31 + 16 = 82 points received
   STEP 3: Finally, divide the points received by the possible points to get the percentage of the total.
   82 ÷ 100 = 82%

2. Two people are going to give money to a foundation for a project. Person A will provide one-half of the money. Person B will donate one-eighth of the money. What fraction represents the unfunded portion of the project?
   A. $1/16$
   B. $1/8$
   C. $1/4$
   D. $5/8$
   E. $3/8$

3. What is the lowest common denominator for the following equation?
   $$\left(\frac{1}{3}+\frac{11}{5}\right)+\left(\frac{1}{15}-\frac{4}{5}\right)$$
   A. 3
   B. 5
   C. 15

D. 45
E. 75

> Questions 2 and 3 are computation and problem solving questions on performing calculations on fractions.

Tips and Explanations:

2. The correct answer is E.
You will see practical problems involving fractions like this one on the exam.
The sum of all contributions must be equal to 100%, simplified to 1. STEP 1: Set up an equation. Let's say that the variable U represents the unfunded portion of the project.
So, the equation that represents this problem is $A + B + U = 1$
STEP 2: Substitute with the fractions that have been provided.

$$\frac{1}{2} + \frac{1}{8} + U = 1$$

STEP 3: Find the lowest common denominator (LCD). Finding the lowest common denominator means that you have to make all of the numbers on the bottoms of the fractions the same. Remember that you need to find the common factors of the denominators in order to find the LCD.
We know that 2 and 4 are factors of 8 because 2 × 4 = 8.
So, the LCD for this question is 8 since the denominator of the first fraction is 2 and because 2 is a factor of 8.
STEP 4: Convert the fractions into the lowest common denominator to solve the problem. We put the fractions into the LCD as follows:

$$\frac{1}{2} + \frac{1}{8} + U = 1$$

$$\left(\frac{1}{2} \times \frac{4}{4}\right) + \frac{1}{8} + U = 1$$

$$\frac{4}{8} + \frac{1}{8} + U = 1$$

$$\frac{5}{8} + U = 1$$

$$\frac{5}{8} - \frac{5}{8} + U = 1 - \frac{5}{8}$$

$$U = 1 - \frac{5}{8}$$

$$U = \frac{8}{8} - \frac{5}{8}$$

$$U = \frac{3}{8}$$

3. The correct answer is C.
We have to find the lowest common denominator (LCD) of the fractions. The LCD for this question is 15. We know this because the product of the other denominators is 3 times 5, which is 15.
We can illustrate the solution as follows:

$$\left(\frac{1}{3}+\frac{11}{5}\right)+\left(\frac{1}{15}-\frac{4}{5}\right)=$$

$$\left[\left(\frac{1}{3}\times\frac{5}{5}\right)+\left(\frac{11}{5}\times\frac{3}{3}\right)\right]+\left[\frac{1}{15}-\left(\frac{4}{5}\times\frac{3}{3}\right)\right]=$$

$$\frac{5}{15}+\frac{33}{15}+\frac{1}{15}-\frac{12}{15}$$

4. A hockey team had 50 games this season and lost 20 percent of them. How many games did the team win?
   A. 8
   B. 10
   C. 20
   D. 18
   E. 40

> Question 4 is a computation and problem solving question on performing calculations on percentages.

Tips and Explanations:

4. The correct answer is E.
For practical problems like this, you must first determine the percentage and formula that you need in order to solve the problem.
Then, you must do long multiplication to determine how many games the team won.
Be careful. The question tells you the percentage of games the team lost, not won.
STEP 1: First of all, we have to calculate the percentage of games won.
If the team lost 20 percent of the games, we know that the team won the remaining 80 percent.
STEP 2: Now do the long multiplication.
   50 games in total
× .80 percentage of games won (in decimal form)
40.0 total games won

5. Carmen wanted to find the average of the five tests she has taken this semester. However, she erroneously divided the total points from the five tests by 4, which gave her a result of 90. What is the correct average of her five tests?
   A. 64
   B. 72
   C. 80
   D. 90
   E. 110

> Question 5 is a question on statistical principles that involves performing arithmetic on basic data related to test scores and determining averages.

Tips and Explanations:

5. The correct answer is B.
   STEP 1: First you need to find the total points that the student earned. You do this by taking Carmen's erroneous average times 4.
   $4 \times 90 = 360$
   STEP 2: Then you need to divide the total points earned by the correct number of tests in order to get the correct average.
   $360 \div 5 = 72$

6. Estimate the result of the following: $502 \div 49.1$
   A. 8
   B. 9
   C. 10
   D. 11
   E. 12

> Question 6 is an estimation and measurement question that requires you to estimate the result of a problem involving division without doing the full computation.

Tips and Explanations:

6. The correct answer is C.
   STEP 1: When doing estimation problems, you need to round the numbers up or down.
   As a rule of thumb, numbers less than 5 will be rounded down to the nearest 0 and numbers of 5 or more will be rounded up to the nearest 10.
   Our problem was $502 \div 49.1$
   So, 502 is rounded down to 500 and 49.1 is rounded up to 50.
   STEP 2: To estimate the result, we then perform the operation on the rounded figures.
   $500 \div 50 = 10$

7. Which of the following is the greatest?
   A. 0.540
   B. 0.054
   C. 0.045
   D. 0.5045
   E. 0.0054

> Question 7 is a numerical and graphic relationship question that requires you to determine if a given number is less than or greater than other numbers. In this problem, the numbers are provided in decimal format.

Tips and Explanations:

7. The correct answer is A.
   Put in extra zeroes and line up the decimal points in a column in order to compare the numbers like this:
   0.5400
   0.0540
   0.0450
   0.5045
   If you are still not sure of your answer, you can remove the decimals as shown below to help you see the answer more clearly.
   5400
   540
   450
   5045
   Therefore, the largest number is .540

8. Which of the following is the most appropriate unit of measurement for the weight of a car?
   A. liters
   B. horsepower
   C. gallons
   D. pounds per square inch
   E. tons

   > Question 8 is an estimation and measurement question on understanding how to measure solid and fluid weights.

Tips and Explanations:

8. The correct answer is E.

   You will need to understand the basic concepts of United States' measurements for the exam. Remember that wet items are measured in pints and quarts, while dry items are measured in ounces and pounds, or tons in the case of extremely heavy quantities.
   Feet and inches are linear measurements; they are not used for weight.
   Liters and gallons are measures of liquid substances. Horsepower measures the strength of an engine.
   Tons measure the weight of heavy items, so it would be suitable for measuring the weight of a car.
   Note that one ton is equal to two thousand pounds.

9. Farmer Brown has a field in which cows craze. He is going to buy fence panels to put up a fence along one side of the field. Each panel is 8 feet 6 inches long. He needs 11 panels to cover the entire side of the field. How long is the field?
   A. 60 feet 6 inches
   B. 72 feet 8 inches
   C. 93 feet 6 inches
   D. 102 feet 8 inches
   E. 110 feet 6 inches

   > Question 9 is an estimation and measurement question on measuring length and perimeter.

Tips and Explanations:

9. The correct answer is C.
Each panel is 8 feet 6 inches long, and he needs 11 panels to cover the entire side of the field. So, we need to multiply 8 feet 6 inches by 11, and then simplify the result.
Step 1: 8 feet × 11 = 88 feet
Step 2: 6 inches × 11 = 66 inches
There are 12 inches in a foot, so we need to determine how many feet and inches there are in 66 inches.
66 inches ÷ 12 = 5 feet 6 inches
Step 3: Now add the two results together.
88 feet + 5 feet 6 inches = 93 feet 6 inches

10. Marta uses one jar of coffee every 6 days. Approximately how many jars of coffee does she use per month?
    A. 2
    B. 3
    C. 5
    D. 6
    E. 7

> Question 10 is a problem solving question on performing calculations on whole numbers.

Tips and Explanations:

10. The correct answer is C.
There are 30 or 31 days in most months, so we need to take the number of days that it takes Marta to use one jar of coffee (which is 6 days in this problem) and divide that number into 30.
30 ÷ 6 = 5 jars per month

11. Jonathan can run 3 miles in 25 minutes. If he maintains this pace, how long will it take him to run 12 miles?
    A. 1 hour and 15 minutes
    B. 1 hour and 40 minutes
    C. 1 hour and 45 minutes
    D. 3 hours
    E. 5 hours

> Question 11 is another problem solving question. This time, you have to solve a practical problem that involves both division and multiplication.

Tips and Explanations:

11. The correct answer is B.
STEP 1: Look to see what information is common to both the question and to the information provided. Here we have the information that he can run 3 miles in 25 minutes. The question is asking how long it will take him to run 12 miles, so the commonality is miles.

STEP 2: Next, you need to find out how many 3-mile increments there are in 12 miles.
12 ÷ 3 = 4
STEP 3: Then you need to determine the time required to travel the stated distance.
Accordingly, we need to multiply the time for 3 miles by 4.
25 minutes × 4 = 100
So, 100 minutes are needed to run 12 miles.
STEP 4: Finally, simplify into hours and minutes based on the fact that there are 60 minutes in one hour.
100 minutes = 1 hour 40 minutes

12. A census shows that 1,008,942 people live in New Town, and 709,002 people live in Old Town. Which of the following numbers is the best estimate of how many more people live in New Town than in Old Town?
    A. 330,000
    B. 300,000
    C. 33,000
    D. 30,000
    E. 3,000

> Question 12 is another estimation and measurement problem involving estimating the results of problems without doing the full computation. To solve question 12, we need to perform subtraction.

Tips and Explanations:

12. The correct answer is B.
    As stated above, this is another type of estimation question.
    The problem tells us that 1,008,942 people live in New Town, and 709,002 people live in Old Town.
    STEP 1: We need to round the numbers up or down to the nearest thousand as needed.
    1,008,942 is rounded to 1,009,000
    709,002 is rounded to 709,000
    STEP 2: Then subtract the second figure from the first figure in order to get your result.
    1,009,000 – 709,000 = 300,000

13. Anne has taken a standardized college entrance exam. Use the report of her test scores below to answer the question that follows.

Raw Score Part 1	Raw Score Part 2	Mean	Standard Deviation	Percentile
180	230	205	15	78

    Which of the following is a correct interpretation of the score report given above?
    A. Ann scored as well as or better than 78% of the test takers.
    B. Ann scored as well as or better than 85% of the test takers.
    C. 15% of the test takers scored higher than Ann.

D. Ann answered 205 of the questions correctly.
E. Ann will perform well at college.

> Question 13 is a statistical principles problem on interpreting standardized test scores and understanding how a particular student has performed relative to other students.

Tips and Explanations:

13. The correct answer is A.
    The raw score represents the number of questions that were answered correctly.
    The mean is the average of the first two raw scores. In other words, we can calculate the mean like this: (180 + 230) ÷ 2 = 205
    Standard deviation measures the variation from the mean or average.
    The percentile rank of a score is the percentage of test-takers that scored the same or lower than the student in question. For instance, a percentile score of 60 means that 60% of the test-takers scored the same or lower than a particular student.
    In our question, Anne's scores were in the 78th percentile, so Ann scored as well as or better than 78% of the test takers.

14. Simplify the following mathematical expression: −183 + 56 + (−17)
    A. 110
    B. 144
    C. −110
    D. −144
    E. −256

> Question 14 is a computation problem involving both positive and negative numbers.

Tips and Explanations:

14. The correct answer is D.
    Be careful with negative signs when answering questions like this one. You might want to add the negatives together before adding in the positive numbers.
    −183 + 56 + (−17) = ?
    STEP 1: Deal with the negative numbers.
    −183 − 17 = −200
    STEP 2: Add in the positive number by reducing it from the negative number.
    −200 + 56 = −144

15. In a high school, 17 out of every 20 students participate in a sport. If there are 800 students at the high school, what is the total number of students that participate in a sport?
    A. 120 students
    B. 640 students
    C. 680 students
    D. 776 students
    E. 780 students

> Question 15 is another problem solving question. You will have to identify the required facts in the problem and then perform both division and multiplication in order to find the solution.

Tips and Explanations:

15. The correct answer is C.
    Remember that for questions like this one, you have to find the commonality between the facts in the question and the requested information for the solution.
    In this question, the commonality is the number of students.
    The question tells us that 17 out of every 20 students participate in a sport and that there are 800 total students.
    STEP 1: Determine how many groups of 20 can be formed from the total of 800.
    800 ÷ 20 = 40 groups of 20 students in the school
    STEP 2: To solve the problem, you then need to multiply the number of participants per group by the possible number of groups.
    In this problem, there are 17 participants per every group of 20.
    There are 40 groups of 20.
    So, we multiply 17 by 40 to get our answer.
    17 × 40 = 680 students

16. The owner of a carnival attraction launches toy boats of different colors. There are 15 boats in total: 5 are blue, 3 are red, and 7 are green. If the first boat launched is green, and the next boat launched is selected at random, what is the probability that the next boat is blue? Note that a boat cannot be launched more than one time.
    A. $5/14$
    B. $6/14$
    C. $5/15$
    D. $4/15$
    E. $10/15$

    > Question 16 is a statistical principles problem that involves understanding basic probability in order to make predictions based on the data provided.

Tips and Explanations:

16. The correct answer is A.
    For questions on probability like this one, you need to reduce the quantity of the total data set by the quantity of items used.
    STEP 1: Determine the total amount in the data set before any items are removed.
    There are 15 boats in total before the first boat is launched.
    5 + 3 + 7 = 15
    STEP 2: Determine the numbers of items in the data set after items have been removed.
    One boat is launched, so the amount in the data set is now 14.
    In other words, the new data set is 14 since 15 − 1 = 14
    STEP 3: Determine the amount in the subset.
    The first boat launched is green, so we need to reduce the subset for that particular color by 1.
    The new total for the green subset becomes 6 since 7 − 1 = 6

However, the question is asking us about the probability that the second boat will be blue, not green. There are 5 blue boats, and a blue boat has not been launched so far.
STEP 4: The probability is expressed as a fraction.
The amount in the subset (5 blue boats) goes on the top of the fraction and the amount of items left in the data set (14 boats left) goes on the bottom.
So, the answer is $5/14$.

17. A new skyscraper is being erected in the city center. The foundation of the building extends 1,135 feet below ground. The building itself, when erected, will measure 13,975 feet above ground. Which of the following is the best estimate of the distance between the deepest point of the foundation below ground and the top of the erected building above ground?
    A. 12,000 feet
    B. 13,000 feet
    C. 14,000 feet
    D. 15,000 feet
    E. 16,000 feet

> Question 17 is an estimation problem that involves a measurement below ground and another measurement above ground.

Tips and Explanations:

17. The correct answer is D.
    The foundation of the building extends 1,135 feet below ground. The building itself, when erected, will measure 13,975 feet above ground. STEP 1: Look at the answer options in order to determine what increments are required.
    Here, we see that the answer options are in increments of one thousand.
    STEP 2: Perform the rounding on both figures.
    So, we round each number up or down to the nearest thousand.
    1,135 is rounded down to 1,000
    13,975 is rounded up to 14,000
    STEP 3: Perform the necessary mathematical computation.
    We add the two figures together from above in order to get our result.
    1,000 + 14,000 = 15,000

18. Mrs. Johnson is going to give candy to the students in her class. The first bag of candy that she has contains 43 pieces. The second contains 28 pieces, and the third contains 31 pieces. If there are 34 students in Mrs. Johnson's class, and the candy is divided equally among all of the students, how many pieces of candy will each student receive?
    A. 3 pieces
    B. 4 pieces
    C. 5 pieces
    D. 51 pieces
    E. 102 pieces

> Question 18 is a computation and problem solving question that involves both addition and division.

Tips and Explanations:

18. The correct answer is A.
    STEP 1: First of all, we need to find out how many pieces of candy there are in total.
    43 + 28 + 31 = 102 total pieces of candy
    STEP 2: We need to divide the total amount of candy by the number of students in order to find out how much candy each student will get.
    102 total pieces of candy ÷ 34 students = 3 pieces of candy per student

19. One hundred students took an English test. The 55 female students in the class had an average score of 87, while the 45 male students in the class had an average of 80. What is the average test score for all 100 students in the class?
    A. 82.00
    B. 83.15
    C. 83.50
    D. 83.85
    E. 84.00

    > Question 19 is another statistical principles problem on performing operations on data relating to test scores and calculating averages.

Tips and Explanations:

19. The correct answer is D.
    STEP 1: First of all, you have to calculate the total amount of points earned by the entire class. Multiply the female average by the amount of female students.
    Total points for female students: 87 × 55 = 4785
    Then multiply the male average by the amount of male students.
    Total points for male students: 80 × 45 = 3600
    STEP 2: Then add these two amounts together to find out the total points scored by the entire class.
    Total points for entire class: 4785 + 3600 = 8385
    STEP 3: When you have calculated the total amount of points for the entire class, you divide this by the total number of students in the class to get the class average. There are 100 total students in this class because there are 55 females and 45 males.
    8385 ÷ 100 = 83.85

20. Two people are going to work on a job. The first person will be paid $7.25 per hour. The second person will be paid $10.50 per hour. A represents the number of hours the first person will work, and B represents the number of hours the second person will work. What equation represents the total cost of the wages for this job?
    A. 17.75AB
    B. 17.75 ÷ AB
    C. AB ÷ 17.75
    D. (7.25A + 10.50B)
    E. (10.50A + 7.25B)

    > Question 20 is a computation and problem solving question involving the use of an algebraic equation in order to solve a practical problem.

Tips and Explanations:

20. The correct answer is D.
    STEP 1: Assign variables as necessary. The two people are working at different costs per hour, so each person needs to be assigned a variable. A is for the number of hours for the first person, and B is for the number of hours for the second person.
    STEP 2: The cost for each person is calculated by taking the number of hours that the person works by the hourly wage for that person.
    So, the equation for wages for the first person is (7.25 × A)
    The equation for the wages for the second person is (10.50 × B)
    STEP 3: The total cost of the wages for this job is the sum of the wages of these two people.
    (7.25 × A) + (10.50 × B) =
    (7.25A + 10.50B)

21. If $5x - 4(x + 2) = -2$, then $x = ?$
    A. 0
    B. 8
    C. 6
    D. −8
    E. −6

> Question 21 is a computation problem on solving algebraic equations with one unknown variable.

Tips and Explanations:

21. The correct answer is C.
    In order to solve algebraic equations with one unknown variable, you have to multiply and then isolate the x variable
    STEP 1: Perform the multiplication on the parenthetical expression.
    $5x - 4(x + 2) = -2$
    $5x - 4x - 8 = -2$
    STEP 2: Then perform any other operations, such as addition or subtraction.
    $(5x - 4x) - 8 = -2$
    $x - 8 = -2$
    STEP 3: Deal with the remaining whole number.
    $x - 8 = -2$
    $x - 8 + 8 = -2 + 8$
    STEP 4: Isolate the variable to solve the problem
    $x - 8 + 8 = -2 + 8$
    $x - 0 = -2 + 8$
    $x = -2 + 8$
    $x = 6$

22. The Johnsons have decided to remodel their upstairs. They currently have 4 rooms upstairs that measure 10 feet by 10 feet each. When they remodel, they will make one large room that will be 20 feet by 10 feet and two small rooms that will each be 10 feet by 8 feet. The remaining space is to be allocated to a new bathroom. What are the dimensions of the new bathroom?
    A. 4 feet × 10 feet

B. 8 feet × 10 feet
C. 10 feet × 10 feet
D. 4 feet × 8 feet
E. 8 feet × 8 feet

> Question 22 is a more advanced measurement problem that involves length, width, and total linear dimensions.

Tips and Explanations:

22. The correct answer is A.
STEP 1: First, we have to calculate the total square footage available. If there are 4 rooms which are 10 by 10 each, we have this equation:
4 × (10 × 10) = 400 square feet in total
STEP 2: Now calculate the square footage of the new rooms.
20 × 10 = 200
2 rooms × (10 × 8) = 160
200 + 160 = 360 total square feet for the new rooms
STEP 3: The remaining square footage for the bathroom is calculated by taking the total minus the square footage of the new rooms.
400 − 360 = 40 square feet
Since each existing room is 10 feet long, we know that the new bathroom also needs to be 10 feet long in order to fit in. So, the new bathroom is 4 feet × 10 feet.

23. Use the table below to answer the following question:

Sunday	Monday	Tuesday	Wednesday	Thursday	Friday	Saturday
−10°F	−9°F	1°F	6°F	8°F	13°F	12°F

The weather forecast for the coming week is given in the table above. What is the difference between the highest and lowest forecasted temperatures for the week?
A. −2°F
B. −3°F
C. 3°F
D. 22°F
E. 23°F

> Question 23 is a numerical relationship problem that involves identifying and using information from a chart.

Tips and Explanations:

23. The correct answer is E.
The lowest temperature is −10°F, and the highest temperature is 13°F. The difference between these two figures is calculated by subtracting. Be careful when you subtract. In particular, remember that when you see two negative signs together, you need to add. In other words, two negatives make a positive: 13 − (−10) = 13 + 10 = 23

24. Acme Packaging uses string to secure their packages prior to shipment. The string is tied around the entire length and entire width of the package, as shown in the following illustration:

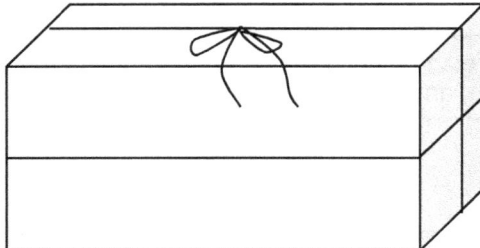

The box is ten inches in height, ten inches in depth, and twenty inches in length. An additional fifteen inches of string is needed to tie a bow on the top of the package. How much string is needed in total in order to tie up the entire package, including making the bow on the top?
A. 55 inches
B. 95 inches
C. 120 inches
D. 130 inches
E. 135 inches

> Question 24 is an advanced measurement problem with a calculation involving length, width, and depth in inches.

Tips and Explanations:

24. The correct answer is E.
For questions that ask you about tying string around a package, you will need to consider the length, width, and depth of the package when doing your calculation.
STEP 1: The string that goes around the top, bottom, and ends of the package will be measured as follows: 20 + 10 + 20 + 10 = 60 inches
STEP 2: The string that goes around the front and back sides and the ends of the package will be calculated similarly since the front and back sides are of the course the same length as the top and bottom.
20 + 10 + 20 + 10 = 60 inches
STEP 3: Don't forget that an additional fifteen inches of rope is needed to tie a bow on the top of the package.
STEP 4: We add these three amounts together to get our total.
60 + 60 + 15 = 135 inches

25. Yesterday a train traveled $117^3/_4$ miles. Today it traveled $102^1/_6$ miles. What is the difference between the distance traveled today and yesterday?
A. 15 miles
B. $15^1/_4$ miles
C. $15^7/_{12}$ miles
D. $15^9/_{12}$ miles
E. $16^5/_6$ miles

> Question 25 is a problem solving question that involves performing calculations on whole numbers and fractions.

Tips and Explanations:

25. The correct answer is C.

    Yesterday the train traveled $117\frac{3}{4}$ miles, and today it traveled $102\frac{1}{6}$ miles. To find the difference, we subtract these two amounts. Because the fraction on the first mixed number is greater than the fraction on the second mixed number, we can subtract the whole numbers and the fractions separately.

    $117\frac{3}{4}$ miles − $102\frac{1}{6}$ miles = ?

    STEP 1: Subtract the whole numbers.

    117 − 102 = 15 miles

    STEP 2: Perform the operation on the fractions by finding the lowest common denominator.

    $\frac{3}{4}$ miles − $\frac{1}{6}$ miles = ?

    In order to find the LCD, we would normally need to find the common factors first.

    Our denominators in this problem are 4 and 6.

    The factors of 4 are:

    1 × 4 = 4

    2 × 2 = 4

    The factors of 6 are:

    1 × 6 = 6

    2 × 3 = 6

    We do not have two factors in common, so we know that we need to find a new denominator which is greater than 6. In this problem, the LCD is 12 since 3 × 4 = 12 and 2 × 6 = 12.

    So, we express the fractions $\frac{3}{4}$ miles + $\frac{1}{6}$ miles from above in their LCD form.

    $\frac{3}{4} \times \frac{3}{3} = \frac{9}{12}$

    $\frac{1}{6} \times \frac{2}{2} = \frac{2}{12}$

    Then subtract these two fractions: $\frac{9}{12} - \frac{2}{12} = \frac{7}{12}$

    STEP 3: Combine the results from the two previous steps to solve the problem.

    $117\frac{3}{4}$ miles − $102\frac{1}{6}$ miles = $15\frac{7}{12}$ miles

26. Liz wants to put new vinyl flooring in her kitchen. She will buy the flooring in square pieces that measure 1 square foot each. The entire room is 8 feet by 12 feet. The cupboards are two feet deep from front to back. Flooring will not be put under the cupboards. A diagram of her kitchen is provided.

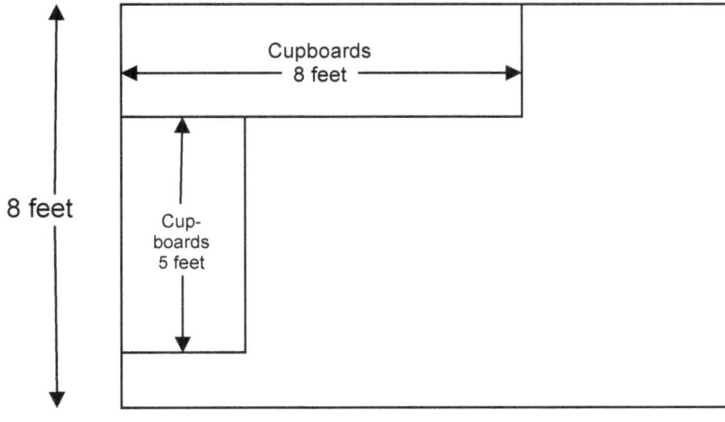

How many pieces of vinyl will Liz need to cover her floor?

A. 84
B. 70
C. 88
D. 96
E. 120

> Question 26 is another advanced measurement problem. Be sure to read the facts provided in problems like this one very carefully.

Tips and Explanations:

26. The correct answer is B.

    For problems like this one, find your solution for each part of the floor and then add these parts together.

    STEP 1: First we will find the square footage of the shaded area below.

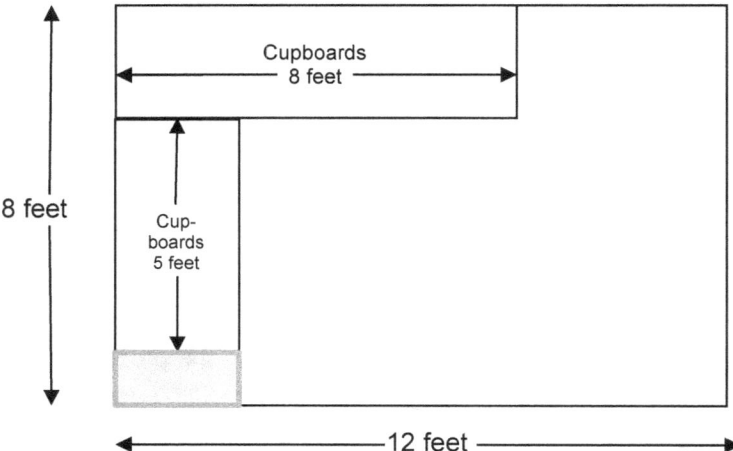

This problem is tricky because flooring is not being placed under the cupboards, so we have to find the square footage for an irregular area.

The cupboards are two feet deep, and the room is 8 feet along the side, so there is a remaining area along the cupboard here of 1 foot (8 feet minus 2 feet for the back cupboard minus 5 feet for the side cupboard = 1 foot) by 2 feet (since the cupboards are two feet deep).

1 foot × 2 feet = 2 square feet

STEP 2: Now find the square footage along the side of the other cupboard.

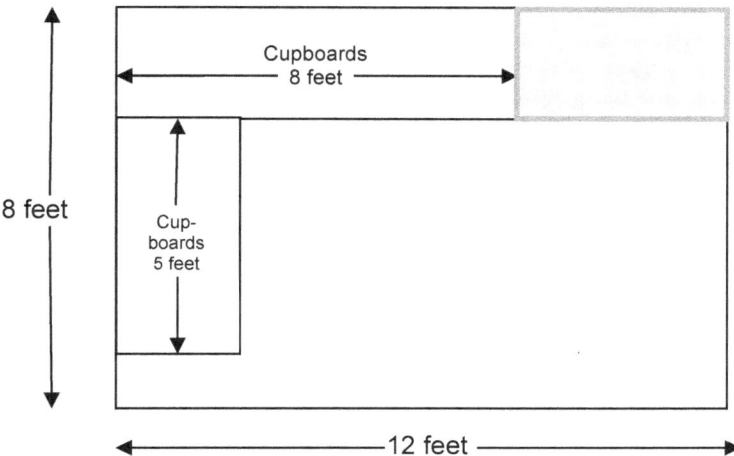

The room is 12 feet along the front, so there is a remaining area along the cupboard here of 4 feet (12 feet minus 8 feet for the length of the back cupboard = 4 feet) by 2 feet (since the cupboards are two feet deep). 4 feet × 2 feet = 8 square feet

STEP 3: Find the square footage for the remaining floor area.

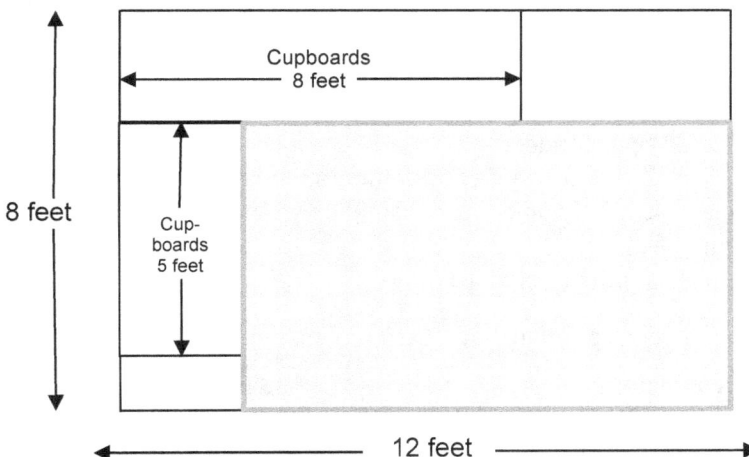

The room is 12 feet along the front and the cupboards are two feet deep, so there is a remaining floor area along the front of 10 feet (12 feet minus 2 feet for the depth of the side cupboard = 10 feet).

The room is 8 feet along the side and the cupboards are two feet deep, so there is a remaining floor area along the side here of 6 feet (8 feet minus 2 feet for the depth of the back cupboard = 10 feet). So the remaining area here is 60 square feet.

10 feet × 6 feet = 60 square feet

STEP 4: Now add the three results together.

Result from step 1: 1 foot × 2 feet = 2 square feet

Result from step 2: 4 feet × 2 feet = 8 square feet
Result from step 3: 10 feet × 6 feet = 60 square feet
2 + 8 + 60 = 70 square feet

27. During each flight, a flight attendant must count the number of passengers on board the aircraft. The morning flight had 52 passengers more than the evening flight, and there were 540 passengers in total on the two flights that day. How many passengers were there on the evening flight?
    A. 244
    B. 296
    C. 488
    D. 540
    E. 592

> Question 27 is a computation and problem solving question that involves allocating more items to one part of the total than to the other part of the total.

Tips and Explanations:

27. The correct answer is A.
    The problem tells us that the morning flight had 52 passengers more than the evening flight, and there were 540 passengers in total on the two flights that day.
    STEP 1: First of all, we need to deduct the difference from the total: 540 – 52 = 488
    In other words, there were 488 passengers on both flights combined, plus the 52 additional passengers on the morning flight.
    STEP 2: Now divide this result by 2 to allocate an amount of passengers to each flight.
    488 ÷ 2 = 244 passengers on the evening flight
    So, the evening flight had 244 passengers
    Had the question asked you for the amount of passengers on the morning flight, you would have had to add back the amount of additional passengers to find the total amount of passengers for the morning flight.
    244 + 52 = 296 passengers on the morning flight

28. Sam is driving a truck at 70 miles per hour. At 10:30 am, he sees this sign:

    | Brownsville | 35 miles |
    | Dunnstun | 70 miles |
    | Farnam | 140 miles |
    | Georgetown | 210 miles |

    After Sam sees the sign, he continues to drive at the same speed. At 11:00 am, how far will he be from Farnam?
    A. He will be in Farnam.
    B. He will be 35 miles from Farnam.
    C. He will be 70 miles from Farnam.
    D. He will be 105 miles from Farnam.
    E. He will be 175 miles from Farnam.

    > Question 28 is a practical problem solving question on calculating distance traveled.

Tips and Explanations:

28. The correct answer is D.
    Sam is driving at 70 miles per hour, and at 10:30 am he is 140 miles from Farnam.
    STEP 1: We need to find out how far he will be from Farnam at 11:00 am, so we need to work out how far he will travel in 30 minutes.
    STEP 2: If Sam is traveling at 70 miles an hour, then he travels 35 minutes in half an hour.
    70 miles in one hour × $\frac{1}{2}$ hour = 35 miles
    STEP 3: If he was 140 miles from Farnam at 10:30 am, he will be 105 miles from Farnam at 11:00 am.
    140 – 35 = 105 miles

29. In a math class, $\frac{1}{3}$ of the students fail a test. If twelve students have failed the test, how many students are in the class in total?
    A. 15
    B. 16
    C. 36
    D. 38
    E. 48

    > Question 29 is an advanced practical problem solving question involving the use of fractions.

Tips and Explanations:

29. The correct answer is C.
    The twelve students who failed the test represent one-third of the class. Since one-third of the students have failed, we can think of the class as being divided into three groups:
    Group 1: The 12 students who failed
    Group 2: 12 students who would have passed
    Group 3: 12 more students who would have passed
    So, the class consists of 36 students in total.
    In other words, we need to multiply by three to find the total number of students.
    12 × 3 = 36

30. Mark owns a bargain bookstore that sells every book for $5. Last week, his sales were $525. This week his sales figure was $600. How many more books did Mark sell this week, compared to last week?
    A. 5
    B. 15
    C. 25
    D. 75
    E. 105

    > Question 30 is a practical problem involving the calculation of the number of units sold.

Tips and Explanations:

30. The correct answer is B.
    The problem tells us that sales this week were $600 and sales last week were $525.

STEP 1: First, we need to find the difference in sales between the two weeks.
$600 - $525 = $75 more in sales this week
STEP 2: Since each book is sold for $5, we divide this figure into the total in order to find out how many books were sold.
$75 more sales ÷ $5 per book = 15 more books sold this week

31. A candy store sells chocolate candy bars. At the beginning of the day, it has 60 candy bars available for sale in total. 25 of them are milk chocolate, 20 are white chocolate, and 15 are dark chocolate. At the close of business that day, 7 candy bars have not been sold. Which of the following details can be determined from the information given above?
    A. average sales of candy bars that week
    B. amount of white dark chocolate bars sold that day
    C. total amount of candy bars sold that day
    D. which candy bar is the most popular normally
    E. the difference between the amount of milk chocolate and dark chocolate candy bars sold that day

> Question 31 is a computation and problem solving question on determining if enough information is provided in order to solve a problem.

Tips and Explanations:

31. The correct answer is C.
    For problems like this, you will need to read the information provided in the problem carefully and then rule out the incorrect answers one by one.
    The problem does not tell us the amount of candy bars sold on other days of the week or year, so we cannot calculate an average for the week as stipulated in answer choice A, nor can we determine which candy bar is the most popular normally, as stipulated in answer choice D.
    The problem also does not tell us the amount of each type of chocolate bar sold. We do not know how many of the candy bars sold were milk chocolate, nor do we know how many were white or dark chocolate. Therefore, we cannot determine the information requested in answer choices B and E.
    We can determine the amount of candy bars sold that day as stated in answer choice C since we know that there were 60 candy bars at the beginning of the day and 7 have not been sold, so 53 candy bars were sold that day.

32. The price of socks is $2 per pair and the price of shoes is $25 per pair. Anna went shopping for socks and shoes, and she paid $85 in total. In this purchase, she bought 3 pairs of shoes. How many pairs of socks did she buy?
    A. 2
    B. 3
    C. 5
    D. 8
    E. 15

> Question 32 is a practical problem solving question involving calculations on prices per unit.

Tips and Explanations:

32. The correct answer is C.
Assign a different variable to each item, and then make an equation by multiplying each variable by its price.
STEP 1: Assign the variables.
Let's say that the number of pairs of socks is S and the number of pairs of shoes is H.
STEP 2: Set up your equation.
Your equation is: $(S \times \$2) + (H \times \$25) = \$85$
STEP 3: We know that the number of pairs of shoes is 3, so put that in the equation and solve it.
$(S \times \$2) + (H \times \$25) = \$85$
$(S \times \$2) + (3 \times \$25) = \$85$
$(S \times \$2) + \$75 = \$85$
$(S \times \$2) + 75 - 75 = \$85 - \$75$
$(S \times \$2) = \$10$
$\$2S = \$10$
$\$2S \div 2 = \$10 \div 2$
$S = 5$

33. Which of the following mathematical expressions equals $3/xy$?
    A. $3/x \times 3/y$
    B. $3 \div 3xy$
    C. $3 \div (xy)$
    D. $1/3 \div 3xy$
    E. $1/3 \div (x3y)$

> Question 33 is a numerical relationship problem on identifying mathematical equivalents.

Tips and Explanations:

33. The correct answer is C.
When you see a fraction, the line in the fraction can be treated as the division symbol. For example, $3/5 = 3 \div 5$
Using the same principle, $3/xy = 3 \div (xy)$

34. Which of the following numbers is between 4,789,321 and 4,901,312?
    A. 4,587,624
    B. 4,780,201
    C. 4,789,231
    D. 4,789,320
    E. 4,792,558

> Questions 34 and 35 are numerical relationship problems on determining if a given number is less than or greater than other numbers.

Tips and Explanations:

34. The correct answer is E.

    For questions like this one, you can line up all of the seven figures in a column in order to compare them.

       4,789,321
    A. 4,587,624
    B. 4,780,201
    C. 4,789,231
    D. 4,789,320
    E. 4,792,558
       4,901,312

    Comparing each digit of each figure as we go along, we can see that answer choices A to D are each less than 4,789,321. Answer choice E is greater than 4,789,321 and less than 4,901,312, so it is correct.

35. If the value of x is between 0.0007 and 0.0021, which of the following could be x?
    A. 0.0012
    B. 0.0006
    C. 0.0022
    D. 0.022
    E. 0.08

35. The correct answer is A.

    This problem is like the previous one, except this question involves decimals. For problems with decimals, line the figures up in a column and add zeroes to fill in the column as shown.

       0.0007
    A. 0.0012
    B. 0.0006
    C. 0.0022
    D. 0.0220
    E. 0.0800
       0.0021

    Answer choice B is less than 0.0007, and answer choices C, D, and E are greater than 0.0021. Answer choice A (0.0012) is between 0.0007 and 0.0021, so it is the correct answer.

36. Terry runs 9 miles every day. If his daily run is rounded up to the nearest 5 miles, which of the following is the best estimate of how many miles he runs every 5 days?
    A. 25
    B. 35
    C. 45
    D. 50
    E. 70

    > Question 36 is an estimation problem involving the use of rounding in order to find the distance traveled.

Tips and Explanations:

36. The correct answer is D.
    STEP 1: Looking at the answer choices, we can see that we need to round to the nearest increment of 5.
    So, for this problem, think about the increments of 5: 5, 10, 15, 20, 25, etc.
    STEP 2. Perform the rounding.
    9 miles per day is rounded up to 10 miles per day.
    STEP 3: Multiply to find the solution.
    We then multiply this figure by the number of days to get our result.
    10 miles per day × 5 days = 50 miles every five days

37. Kathy is on a diet. During week 1, she lost 1.07 pounds. During week 2, she lost 2.46 pounds. During week 3, she lost 3.92 pounds. If each week's weight loss amount is rounded up or down to the nearest one-tenth of a pound, what is the estimate of Kathy's weight loss for the entire 3 weeks?
    A. 7 pounds
    B. 7.40 pounds
    C. 7.45 pounds
    D. 7.50 pounds
    E. 8 pounds

    > Question 37 is an estimation problem involving the rounding of decimals.

Tips and Explanations:

37. The correct answer is D.
    One-tenth is expressed is decimal form like this: 0.1
    So, any amount that has a decimal less than 0.05 is rounded down and decimals of 0.05 and greater are rounded up.
    STEP 1: We do the rounding as follows:
    1.07 is rounded up to 1.1
    2.46 is rounded up to 2.5
    3.92 is rounded down to 3.9
    STEP 2: Then add the rounded figures together to get your result.
    1.1 + 2.5 + 3.9 = 7.5

38. Use the information provided in the box below to answer the question that follows.

    > - The police station is 10 miles away from the fire station
    > - The fire station is 6 miles away from the hospital.

    Based on the information in the box, what conclusions can be made?
    A. The police station is no more than 6 miles away from the hospital.
    B. The police station is no more than 10 miles away from the hospital.
    C. The police station is exactly 6 miles away from the hospital.
    D. The fire station is exactly 10 miles away from the hospital.
    E. The police station is no more than 16 miles away from the hospital.

    > Question 38 is a numerical and graphic relationship word problem on logical relationships.

Tips and Explanations:

38. The correct answer is E.
   For questions about distance like this one, keep in mind that the locations may or may not lie on a straight line. For example, the locations could be laid out like this:

   Police station ⎯⎯⎯⎯⎯⎯→ Fire station ⎯⎯⎯⎯⎯⎯→ Hospital
   　　　　　　　　10 miles　　　　　　　　6 miles

   In the layout above, the police station would be 16 miles from the hospital.
   However, the locations could also be laid out like this:

   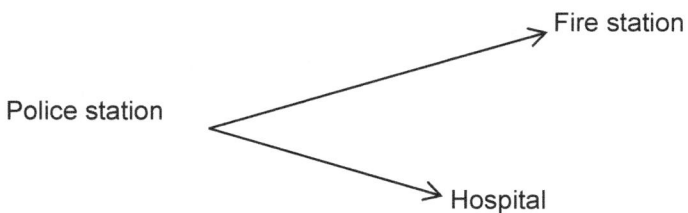

   We can see that the locations will be the farthest from each other if they are laid out on a straight line as in the first example above.
   In other words, a person could always go to the hospital by traveling to the fire station from the police station (10 miles) and then traveling from the fire station to the hospital (6 miles).
   Therefore, the police station would never be more than 16 miles away from the hospital, regardless of the layout.

39. Kieko needs to calculate 16% of 825. Which of the following formulas can she use?
    A. 825 × 16
    B. 16 × 825
    C. 825 × 16
    D. 825 × 1.6
    E. 825 × 0.16

   > Question 39 is a computation and problem solving question on identifying mathematical equivalents of percentages and decimals.

Tips and Explanations:

39. The correct answer is E.
    A percentage can always be expressed as a number with two decimal places.
    For example, 15% = 0.15 and 20% = 0.20
    In our problem, 16% = 0.16
    Therefore, E is the correct answer.

40. Wei Lei bought a shirt on sale. The original price of the shirt was $18, and he got a 40% discount. What was the sales price of the shirt?
    A. $7.20
    B. $10.80
    C. $11.80
    D. $17.28
    E. $17.60

> Question 40 is another practical problem on the calculation of discounts.

Tips and Explanations:

40. The correct answer is B.
    STEP 1: First of all, you need to calculate the amount of the discount.
    $18 original price × 40% =
    $18 × 0.40 = $7.20 discount
    STEP 2: Then deduct the amount of the discount from the original price to calculate the sales price of the item.
    $18 original price - $7.20 discount = $10.80 sales price

41. Professor Smith uses a system of extra-credit points for his class. Extra-credit points can be offset against the points lost on an exam due to incorrect responses. David answered 18 questions incorrectly on the exam and lost 36 points. He then earned 25 extra credit points. By how much was his exam score ultimately lowered?
    A. –11
    B. 11
    C. 18
    D. 25
    E. 36

> Question 41 is another problem on performing arithmetic on data relating to test scores. It involves operations with both positive and negative numbers.

Tips and Explanations:

41. The correct answer is B.
    If David answered 18 questions incorrectly on the exam and lost 36 points, and he then earned 25 extra credit points, his score was lowered by 11 points.
    STEP 1: To do the calculation, we need to take the points lost on the exam and add the extra credit points.
    –36 + 25 = –11
    STEP 2: Since the question is asking how much the score was lowered, you need to give the amount as a positive number.

42. What number is next in this sequence? 2, 4, 8, 16
    A. 18
    B. 20
    C. 24
    D. 32
    E. 36

> Questions on sequences like question 42 are another type of numerical relationship problem.

Tips and Explanations:

42. The correct answer is D.
    Try to find the pattern of relationship between the numbers.

Here, we can see this pattern:

2 × 2 = 4

4 × 2 = 8

8 × 2 = 16

In other words, the next number in the sequence is always double the previous number.

Therefore the answer is: 16 × 2 = 32

43. The county is proposing a 7.5% increase in its annual real estate tax. If the tax is currently $480 per year, how much would the tax be if the proposed increase is approved?
    A. $444
    B. $487
    C. $516
    D. $840
    E. $3330

> Question 43 is a practical problem solving question on using percentages to calculate a tax increase.

Tips and Explanations:

43. The correct answer is C.
    STEP 1: Calculate the amount of the tax increase.
    $480 × 7.5% = ?
    $480 original tax amount × 0.075 = $36 proposed increase in tax
    STEP 2: Then add the increase to the original amount to get the amount of the tax after the proposed increase.
    $480 original tax + $36 increase in tax = $516 tax after increase

44. Which one of the values will correctly satisfy the following mathematical statement:
    $2/3 < ? < 7/9$
    A. $1/3$
    B. $1/5$
    C. $2/6$
    D. $1/2$
    E. $7/10$

> Question 44 is a numerical relationship problem on ordering fractions from least to greatest.

Tips and Explanations:

44. The correct answer is E.
    This is another question involving common denominators.
    The question is: $2/3 < ? < 7/9$
    STEP 1: First of all, we need to find a common denominator for the fractions in the equations, as well as for all of the answer choices. In order to complete the problem quickly, you should not try to find the lowest common denominator, but just find any common denominator.
    We can do this by expressing all of the numbers with a denominator of 90 since 9 is the largest denominator in the equation and 10 is the largest denominator in the answer choices.
    $2/3 × 30/30 = 60/90$
    $7/9 × 10/10 = 70/90$

STEP 2: Then, express the original equation in terms of the common denominator.
$60/90 < ? < 70/90$

STEP 3: Then express the answer choices in terms of the common denominator.
A. $1/3 \times 30/30 = 30/90$
B. $1/5 \times 18/18 = 18/90$
C. $2/6 \times 15/15 = 30/90$
D. $1/2 \times 45/45 = 45/90$
E. $7/10 \times 9/9 = 63/90$

STEP 4: Compare the results to find the answer.
By comparing the numerators (the top numbers of the fractions), we can see that $63/90$ lies between $60/90$ and $70/90$, so E is the correct answer.

45. Use the information in the box below to answer the question that follows.

   - An orchard grows apples for resale.
   - If the apple is 8 inches or more around, it is classified as grade A and sold to exclusive retailers.
   - If the apple measures less than 8 inches around, but more than 4 inches around, it is classified as grade B and sold to wholesalers.
   - If the apple measures 4 inches or less around it is classified as grade C.
   - Apples with a grade C classification are rejected for human consumption and are sold to animal food manufacturers.

   If an apple measures exactly 4 inches around, which of the following statements could be true?
   A. The apple will be classified as grade B.
   B. The apple will be sold to exclusive retailers.
   C. The apple will be sold to wholesalers.
   D. The apple will not be classified as grade C.
   E. The apple will not be eaten by people.

   > Questions 45 and 46 are further word problems on logical relationships.

Tips and Explanations:

45. The correct answer is E.
   Read the facts of problems like this one very carefully. The facts provided in the problem tell us that if an apple measures 4 inches or less around it is classified as grade C, which is sold to animal food manufacturers. The apple in this problem is exactly 4 inches, so it is a grade C apple. Therefore, the apple will not be eaten by people, but by animals.

46. Use the information in the box below to answer the question that follows.

   School will be held every weekday from Monday to Friday from 15th August until 22nd December from 8:30 am to 3:00 pm.
   However, if the temperature is more than 100 degrees, school will be dismissed at 11:30 am. School will not be held on public holidays.

   If Tom did not go to school today, then which of the following statements must be true?
   A. It is the 22nd of August.
   B. The temperature exceeds 100 degrees.
   C. It is a public holiday or the temperature exceeds 100 degrees.

D. It is a public holiday and the temperature exceeds 100 degrees.
E. It is a public holiday or a Saturday or Sunday.

46. The correct answer is E.
This is another type of problem in which you have to assess the available facts, so read carefully and do not make any assumptions that are not supported by the information provided. As stated previously, it is usually best to deal with each answer option one by one for these types of questions.
Answer choice A is incorrect. If it is the 22nd of August, Tom would be in school because school is held every weekday from Monday to Friday from 15th August until 22nd December.
Answer choices B, C, and D are incorrect because even if the temperature was in excess of 100 degrees, Tom would have attended school from 8:30 to 11:30 am.
Therefore, we know that answer E is the correct answer. We also know that E is correct because school is held only on weekdays, and answer E stipulates that it is a Saturday, Sunday, or public holiday.

47. Use the information in the chart below to answer the question that follows.

Fatal Traffic Accidents in Hawaii, Alaska, Texas, and the Other States				
Year	Hawaii	Alaska	Texas	Other States
1994	2,365	987	3,687	52,187
1999	1,987	882	3,522	48,233
2004	1,784	915	3,601	51,505
2009	1,801	899	3,547	50,689
2014	1,621	823	3,623	49,117

How many fatal traffic accidents occurred outside the state of Texas in 2004?
A. 2,699
B. 3,601
C. 51,505
D. 54,204
E. 55,539

> Questions 47 to 50 are graphic relationship problems with tables, line graphs, and pie charts.

Tips and Explanations:

47. The correct answer is D.
The question is asking you for the total of all states, besides Texas, so you need to add together the amounts for Hawaii, Alaska, and the remaining states for the year 2004.
1,784 + 915 + 51,505 = 54,204

48. Use the information in the graph below to answer the question that follows.

Enrollment at Southwestern College

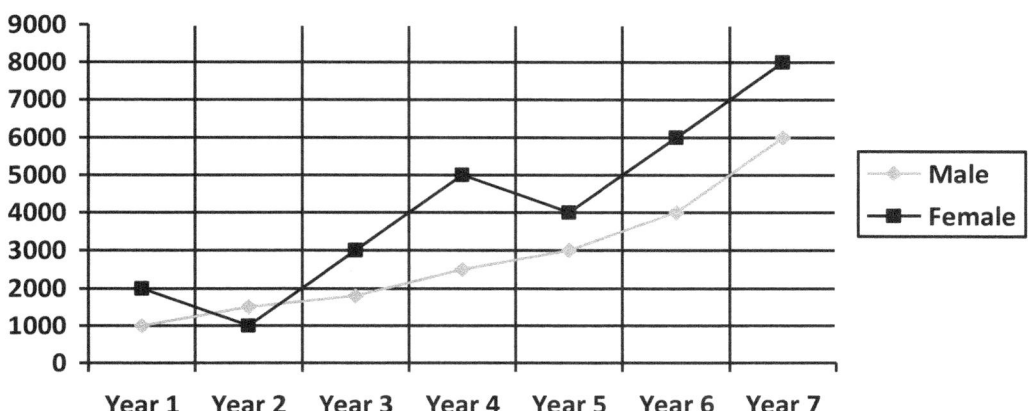

The chart above shows enrollment at Southwestern College over a seven-year period. According to the chart above, what was the largest approximate difference between male and female enrollment in any of the years displayed on the graph?

A. 300
B. 500
C. 1000
D. 2000
E. 2300

48. The correct answer is E.
For graphic questions like this one, you can usually find the answer by visually inspecting the graph. However, be sure to double-check your answer.
The points appear to be the farthest apart in year 4, where we see that the points are more than two lines apart.
The second farthest points are only two lines apart.
Each line represents one thousand students, so the difference between the amount of male and female students in year 4 is approximately 2,300 students.

49. Use the information in the graph below to answer the question that follows.

Favorite Drinks of Customers at Metro Bar and Grill

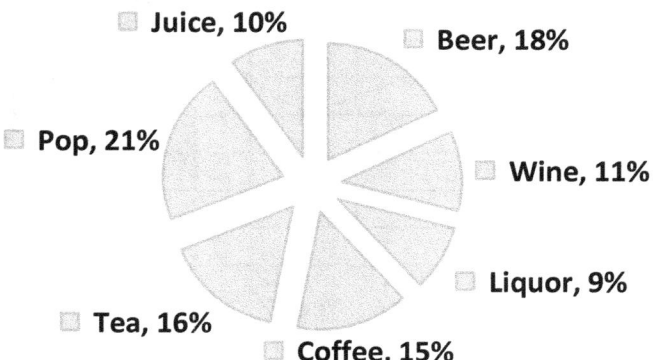

What percentage of customers at Metro Bar and Grill have a favorite drink that is a non-alcoholic beverage?
A. 62%
B. 48%
C. 41%
D. 42%
E. 23%

49. The correct answer is A.
Juice, pop, tea, and coffee are non-alcoholic beverages, so we need to add up these percentages: 10% + 21% + 16% + 15% = 62%

50. Use the information in the graph below to answer the question that follows.

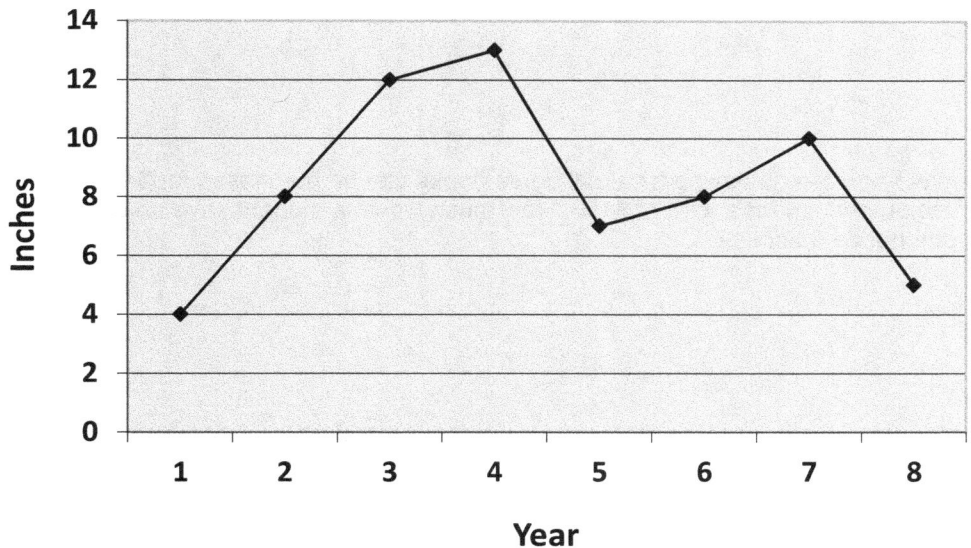

Annual Rainfall in Hudson County

Between what two years did the amount of rainfall change by the greatest amount?

A. between year 1 and year 2
B. between year 2 and year 3
C. between year 4 and year 5
D. between year 5 and year 6
E. between year 7 and year 8

50. The correct answer is C.
The distance between each set of horizontal lines on the graph represents two inches of rain. We can see that the change is biggest between years 4 and 5, since the line of the graph stretches across three horizontal lines between these two years.

CBEST Practice Math Test 2

1. A group of friends are trying to lose weight. Person A lost $14^{3}/_{4}$ pounds. Person B lost $20^{1}/_{5}$ pounds. Person C lost 36.35 pounds. What is the total weight loss for the group?
   A. 70.475
   B. 71.05
   C. 71.15
   D. 71.25
   E. 71.30

2. The university bookstore is having a sale. Course books can be purchased for $40 each, or 5 books can be purchased for a total of $150. How much would a student save on each book if he or she purchased 5 books?
   A. 5
   B. 10
   C. 50
   D. 90
   E. 110

3. 120 students took a math test. The 60 female students in the class had an average score of 95, while the 60 male students in the class had an average of 90. What is the average test score for all 120 students in the class?
   A. 75
   B. 92.5
   C. 93
   D. 93.5
   E. 120

4. Tom bought a shirt on sale for $12. The original price of the shirt was $15. What was the percentage of the discount on the sale?
   A. 2%
   B. 3%
   C. 20%
   D. 25%
   E. 30%

5. A car travels at 60 miles per hour. The car is currently 240 miles from Denver. How long will it take for the car to get to Denver?
   A. 40 minutes
   B. 60 minutes
   C. 4 hours
   D. 5 hours
   E. 6 hours

6. Use the diagram below to answer the question that follows.

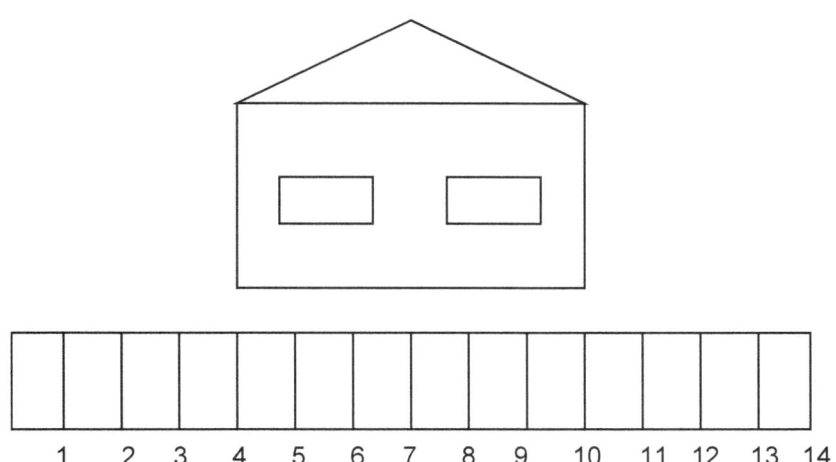

If each rectangle in the ruler below the picture of the house is one unit and the actual length of the house is 36 feet, then what is the scale of the diagram of the house?
A. 1 unit = 6 feet
B. 1 unit = 7.2 feet
C. 1 unit = 9 feet
D. 1 unit = 12 feet
E. 1 unit = 36 feet

7. What is the most appropriate unit of measure for determining the dimensions of a sofa?
A. feet and inches
B. pounds
C. yards
D. ounces
E. tons

8. What is the remainder when 11 is divided by 3?
A. .06
B. .27
C. 2
D. 3
E. 9

9. Mrs. Ramirez is inviting 12 children to her son's birthday party. The children will play pin the tail on the donkey. Mrs. Ramirez has already made 40 tails for the game. She wants to give each child 4 tails to play the game. How many more tails does she need to make?
A. 4
B. 8
C. 10
D. 12
E. 28

10. Yesterday the temperature was 90 degrees. Today it is 10% cooler than yesterday. What is today's temperature?
    A. 80 degrees
    B. 81 degrees
    C. 91 degrees
    D. 99 degrees
    E. 100 degrees

11. A class contains 20 students. On Tuesday 5% of the students were absent. On Wednesday 20% of the students were absent. How many more students were absent on Wednesday than on Tuesday?
    A. 1
    B. 2
    C. 3
    D. 4
    E. 5

12. $1/3 - 1/7 = ?$
    A. $1/21$
    B. $1/4$
    C. $3/7$
    D. $4/21$
    E. $-1/4$

13. Mark's final grade for a course is based on the scores from two tests, A and B. The score from test A counts toward 35% of his final grade. The score from test B counts toward 65% of his final grade. What equation is used to calculate Mark's final grade for this course?
    A. (.65A + .35B)
    B. (.35A + .65B)
    C. (.35A + .65B) ÷ 2
    D. A + B
    E. (A + B) ÷ 2

14. If A represents the number of apples purchased at 20 cents each and B represents the number of bananas purchased at 25 cents each, what equation represents the total value of the purchase?
    A. .45AB
    B. .45 ÷ AB
    C. .20A + .25B
    D. .25A + .20B
    E. .45 + (A × B)

15. If $x - 3 + 5x = 33$, then $x = ?$
    A. 6
    B. 7
    C. 8
    D. 9
    E. 12

16. Pat wants to put wooden trim around the floor of her family room. Each piece of wood is 1 foot in length. The room is rectangular and is 12 feet long and 10 feet wide. How many pieces of wood does Pat need for the entire perimeter of the room?
    A. 22
    B. 44
    C. 100

D. 120
E. 144

17. Ben uses one bag of dog food every 6 days to feed his dog. Approximately how many bags of dog food would Ben require for two months?
    A. 5
    B. 6
    C. 9
    D. 10
    E. 20

18. The snowfall for November is 5 inches less than for December. If the total snowfall for November and December is 35 inches, what was the snowfall for November?
    A. 10 inches
    B. 15 inches
    C. 20 inches
    D. 30 inches
    E. 40 inches

19. Records indicate that there were 12 hospitals in Johnson County in 1995, but this number had increased to 15 hospitals in 2014. There were 12 births on average per hospital in Johnson County in 1995. The total number of births in Johnson County was 240 in 2014. By what amount does the average number of births per hospital in Johnson County for 2014 exceed those for 1995?
    A. 3 births per hospital
    B. 4 births per hospital
    C. 15 births per hospital
    D. 16 births per hospital
    E. 96 births per hospital

20. A magician has a bag of colored scarves for a magic trick that he performs. The bag contains 3 blue scarves, 1 red scarf, 4 green scarves, and 2 orange scarves. If the magician removes scarves at random and the first scarf she removes is red, what is the probability that the next scarf will be orange?
    A. $1/2$
    B. $2/7$
    C. $1/9$
    D. $2/9$
    E. $2/10$

21. Marta can walk one mile in 17 minutes. At this rate, how long would it take her to walk 5 miles?
    A. 1 hour and 5 minutes
    B. 1 hour and 7 minutes
    C. 1 hour and 8 minutes
    D. 1 hour and 15 minutes
    E. 1 hour and 25 minutes

22. Simplify the following expression: −117 + (−25) + 45
    A. −47
    B. −97
    C. −137
    D. 137
    E. 187

23. **Use the information below to answer the question that follows.**

Appleton	Brownsville	Charlestown	Durham	Easton
687 feet below sea level	1586 feet above sea level	253 feet below sea level	542 feet below sea level	1621 feet above sea level

As part of a geography class, students are required to learn the distance above and below sea level of certain towns in their area. What was the difference in feet between the highest and lowest towns in their area according to the above table?
   A. 66 feet
   B. 621 feet
   C. 874 feet
   D. 2273 feet
   E. 2308 feet

24. Clark County had 135,298 cases of infectious disease last year, while Davidson County had 207,121 cases. What number is the best estimate of how many more cases of infectious disease there were in Davidson County?
   A. 12,000
   B. 62,000
   C. 70,000
   D. 71,000
   E. 72,000

25. What is the best estimate of 5,012 × 12?
   A. 50,000
   B. 52,000
   C. 60,000
   D. 70,000
   E. 600,000

26. **Please use the diagram below to answer the question that follows.**

The above diagram depicts a football field. The field is 30 yards wide and 100 yards long. Paint is sprayed on the field in lines that are 10 yards apart, as indicated by the vertical lines in the diagram above. Paint is also sprayed around the entire perimeter of the field. In total, how many yards of paint are sprayed onto the field?
   A. 290
   B. 300
   C. 530
   D. 560
   E. 590

27. Solve for x: 3x – 4 – x = 12
    A. –2
    B. –4
    C. 2
    D. 4
    E. 8

28. A martial arts class has 53 students at the beginning of the year. 15 students have black belts, 22 have brown belts, 8 have blue belts, and 8 have belts of other colors. By the end of the year, 3 of the students with brown belts and 2 of the students with belts of other colors have dropped out of the class. In addition, 4 new students have joined the class.

    Which of the following facts can be determined from the information above?
    A. The total number of students in the class.
    B. The number of students in the class with brown belts.
    C. The number of students in the class with blue belts.
    D. The number of students in the class with black belts.
    E. The number of students in the class with belts of other colors.

29. Sam's final grade for a class is based on his scores from a midterm test (M), a project (P), and a final exam (F). The midterm test counts twice as much as the project, and the final exam counts twice as much as the midterm. Which mathematical expression below can be used to calculate Sam's final grade?
    A. P + M + F
    B. P + M + 2F
    C. P + 2M + F
    D. P + 2M + 2F
    E. P + 2M + 4F

30. Bart is riding his bike at a rate of 12 miles per hour. He arrives in the town of Wilmington at 3:00 pm. The town of Mount Pleasant is 50 miles from Wilmington. How far will Bart be from Mount Pleasant at 5:00 pm if he continues riding his bike at this speed?
    A. 12 miles
    B. 20 miles
    C. 24 miles
    D. 26 miles
    E. 36 miles

31. A ticket office sold 360 more tickets on Friday than it did on Saturday. If the office sold 2570 tickets in total during Friday and Saturday, how many tickets did it sell on Friday?
    A. 360
    B. 1105
    C. 1465
    D. 1565
    E. 2210

32. **Use the chart below to answer the question that follows.**

X	Y
2	4
4	16
6	
8	64
10	100

The chart shows the mathematical relationship between X and Y. What value of Y is missing from the chart?
A. 24
B. 30
C. 32
D. 36
E. 48

33. Tom's height increased by 10% this year. If Tom was 5 feet tall at the beginning of the year, how tall is he now?
A. 5 feet 1 inch
B. 5 feet 5 inches
C. 5 feet 6 inches
D. 5 feet 10 inches
E. 6 feet

34. Read the problem below and answer the question that follows.

> Mary left home at 10:00 am. She drove 150 miles to Newton at a constant rate of 70 miles per hour. She then rested for 30 minutes before driving 105 miles to Lordville. What time was it when she arrived in Lordville?

What piece of information is needed in order to solve the problem?
A. The number of gallons of gas she used for the journey.
B. The speed that Mary traveled from Newton to Lordville.
C. The speed limit on the road from Newton to Lordville.
D. The amount of time she rested in Lordville.
E. Her arrival time at her house on her return journey.

35. Read the problem below and answer the question that follows.

> Dan rode his horse 2 miles to his neighbor's house. It took the horse 15 minutes to make this journey. From his neighbor's house, Dan rode his horse 3 miles into town. What is the average pace of Dan's horse in miles per hour for these two journeys?

What piece of information is needed in order to solve the problem?
A. The distance from Dan's house into town.
B. The amount of time Dan stayed at his neighbor's house.
C. The length of the stride of Dan's horse.
D. Whether Dan's horse trotted or galloped.
E. The amount of time it took to go from the neighbor's house into town.

36. Which mathematical expression is equivalent to 2H + 3H?
A. 5H
B. 5 + H
C. 6H
D. H^5
E. H^6

37. Which of the following numbers is between 2,368,741 and 2,654,802?
A. 2,281,414
B. 2,306,549
C. 2,367,988
D. 2,683,699
E. 2,645,972

38. Beth needs to calculate 15% of 60. Which equation below can she use in order to do so?
    A. 60 × 15
    B. 60 × 1.5
    C. 60 × .15
    D. 60 × .0015
    E. 60 ÷ 15

39. Which of the following mathematical statements is correct?
    A. .001 < .0001 < .00001
    B. .001 < .0010 < .00100
    C. .0001 < .001 < .01
    D. .0010 < .001 < .01
    E. .0100 < .010 < .10

40. $1/3 > ? > 1/9$
    A. $1/2$
    B. $1/4$
    C. $1/10$
    D. $2/3$
    E. $2/5$

41. Carlos buys 2 pairs of jeans for $22.98 each. He later decides to exchange both pairs of jeans for 3 sweaters which cost $15.50 each. Which equation can Carlos use to calculate the extra money he will have to pay for the exchange?
    A. 2 × (22.98 - 15.50)
    B. 3 × (22.98 - 15.50)
    C. (3 × 22.98) − (2 × 15.50)
    D. (3 × 15.50) − (2 × 22.98)
    E. (3 × 15.50) + (2 × 22.98)

42. If the value of x is less than .06 but greater than .006, which one of the following could be x?
    A. .05
    B. .005
    C. .0005
    D. .0006
    E. .00005

43. A museum counts its visitors each day and rounds each daily figure up or down to the nearest 5 people. 104 people visit the museum on Monday, 86 people visit the museum on Tuesday, and 81 people visit the museum on Wednesday. Which figure below best represents the amount of visitors to the museum for the three days, after rounding?
    A. 260
    B. 265
    C. 270
    D. 275
    E. 280

44. Jason does the high jump for his high school track and field team. His first jump is at 3.246 meters. His second is 3.331 meters, and his third is 3.328 meters. If the height of each jump is rounded to the nearest one-hundredth of a meter (also called a centimeter), what is the estimate of the total height for all three jumps combined?
    A. 9.80
    B. 9.89
    C. 9.90

D. 9.91
E. 10.00

45. Use the information below to answer the question that follows.

| • The baseball team practices every Tuesday and Friday. |
| • There will be no practice during the last full week of the month. |
| • There will be no practice in the event of rain. |

If there is practice today, which of the following conclusions can be made?
A. It is the last full week of the month.
B. It is a Tuesday or it is not raining.
C. It is a Tuesday and it is raining.
D. It is a Tuesday or a Friday.
E. It is the last full week of the month or it is raining.

46. Use the table below to answer the question that follows.

| Regional Railway Train Service ||
Departure Time	Arrival Time
9:50 am	10:36 am
11:15 am	12:01 pm
12:30 pm	1:16 pm
2:15 pm	3:01 pm
?	5:51 pm

The journey on the Regional Railway is always exactly the same duration.
What is the missing time in the chart above?
A. 3:30 pm
B. 4:15 pm
C. 4:30 pm
D. 5:05 pm
E. 5:15 pm

47. Use the chart below to answer the question that follows.

| United Kingdom and Entire World Coal Consumption in Tons |||
Year	United Kingdom	Entire World
1925	3528	8741
1945	3679	9523
1965	3598	7413
1985	2565	6528
2005	1201	4889
2010	800	4665

How many tons of coal were consumed outside the United Kingdom in 1965?
A. 3815
B. 3915
C. 4815

D. 6528
E. 7413

48. Use the chart below to answer the question that follows.

Electricity and Gas Consumption for the Past Ten Years

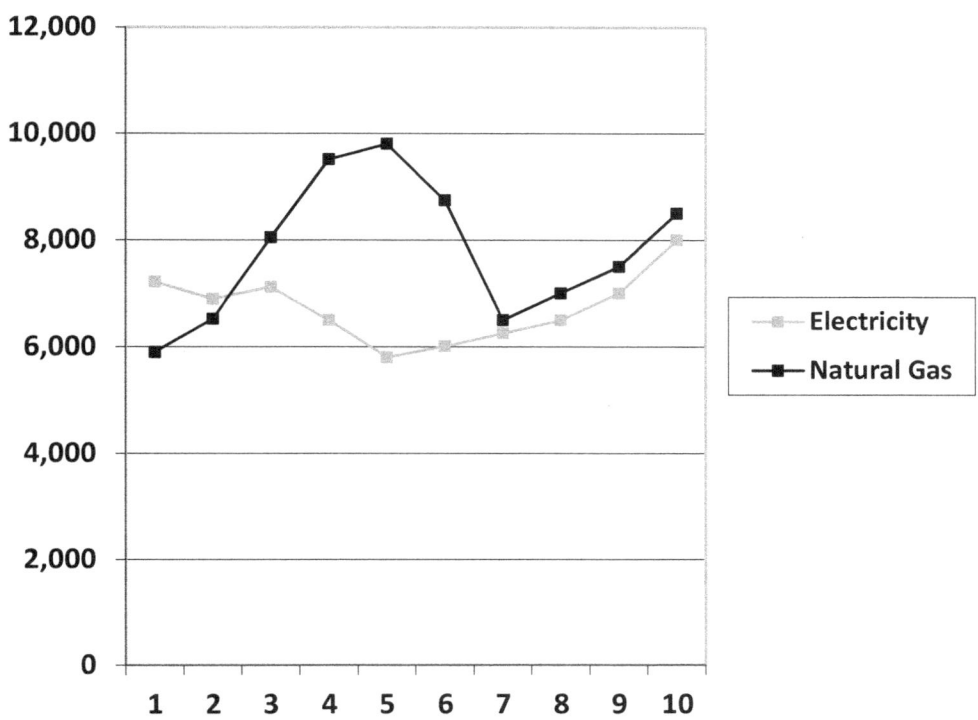

According to the graph, the greatest annual difference in units between the consumption of electricity and the consumption of natural gas occurred in which year?
A. 2,000 units
B. 3,000 units
C. 4,000 units
D. 5,000 units
E. 16,000 units

49. Use the chart below to answer the question that follows.

Crop Production

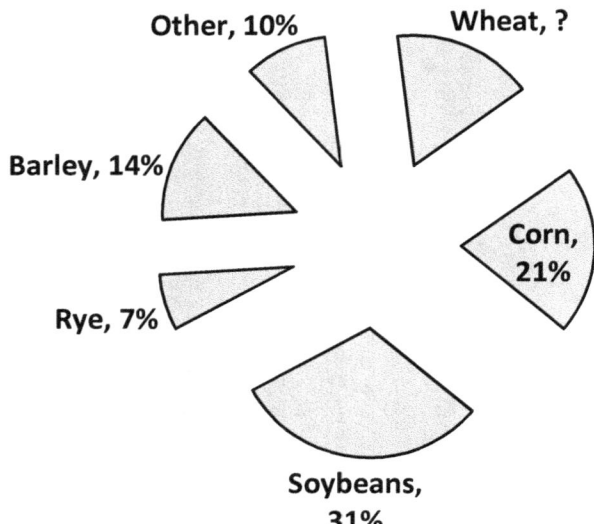

The percentage for wheat production is not provided in the above the chart. What is the percentage of wheat production?
A. 9%
B. 12%
C. 13%
D. 17%
E. 29%

50. Use the graph below to answer the question that follows.

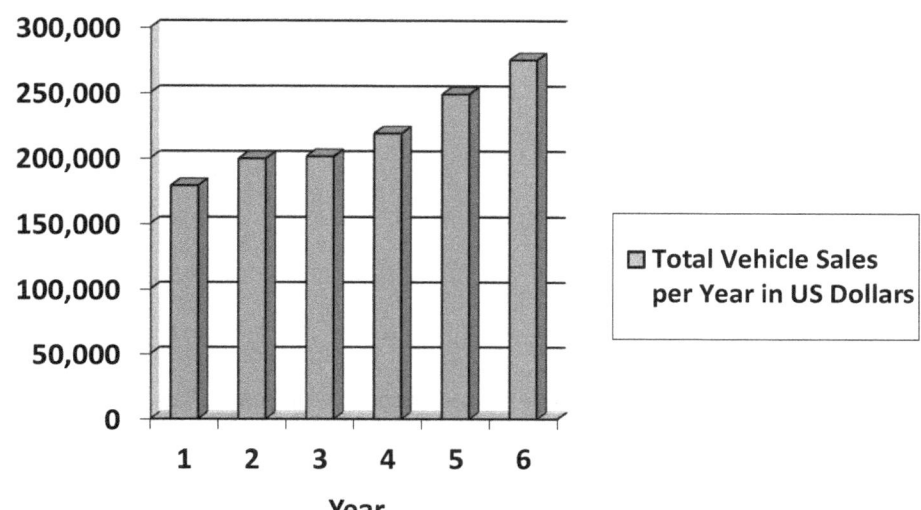

The graph shows sales of vehicles during a six-year period. Between which years did the sales increase the most?
A. Years 1 and 2
B. Years 2 and 3
C. Years 3 and 4
D. Years 4 and 5
E. Years 5 and 6

CBEST Math Practice Test 2 – Answers

1. The correct answer is E.
   We have both fractions and decimals in this problem.
   Convert the fractions in the mixed numbers to decimals.
   $3/4 = 3 ÷ 4 = 0.75$
   $1/5 = 1 ÷ 5 = 0.20$
   Then represent the mixed numbers as decimal numbers.
   Person 1: $14^{3}/_{4}$ = 14.75
   Person 2: $20^{1}/_{5}$ = 20.20
   Person 3: 36.35
   Then add all three amounts together to find the total.
   14.75 + 20.20 + 36.35 = 71.30

2. The correct answer is B.
   First, divide the total price for the multi-purchase by the number of items. In this case, $150 ÷ 5 = $30 for each of the five books.
   Then, subtract this amount from the original price to get your answer.
   $40 – $30 = $10
   Alternatively, you can use the method explained below.
   Calculate the total price for the five books without the discount.
   5 × $40 = $200
   Then subtract the discounted price of $150 from the total.
   $200 - $150 = $50
   Then divide the total savings by the number of books to determine the savings on each book.
   $50 total savings ÷ 5 books = $10 savings per book

3. The correct answer is B.
   You need to find the total points for all the females by multiplying their average by the number of female students. Then do the same to find the total points for all the males.
   Females: 60 × 95 = 5700
   Males: 60 × 90 = 5400
   Then add these two amounts together to get the total for the group.
   5700 + 5400 = 11,100
   Then divide by the total number of students in the class to get your solution.
   11,100 ÷ 120 = 92.5
   So, the correct average is 92.5

4. The correct answer is C.
   Find the dollar amount of the discount first.
   $15 original price – $12 sales price = $3 discount
   Then divide the discount into the original price to get the percentage.
   $3 ÷ $15 = 0.20 = 20%

5. The correct answer is C.
   Divide the miles per hour into the distance left in order to get the time needed.
   240 miles remaining ÷ 60 mph = 4 hours left to travel

6. The correct answer is A.
   Count the number of units that the house spans, rather than trying to subtract units from the total of 14.

If we count the number of units below the house in the drawing, we can see that the house spans 6 units.

Divide this result into the actual length of the house (36 feet) to get the scale of the drawing.

36 feet ÷ 6 units = 6 feet represented by each unit

7. The correct answer is A.

   Pounds, ounces, and tons are used to measure the weight of items, not their dimensions or linear measurements. Feet, inches, and yards are used to measure linear dimensions. Feet and inches are suitable for small items which one usually finds indoors, while yards are used for larger items which one usually finds outdoors. A sofa is an item of furniture which is used inside the house, so "feet and inches" is the best answer.

8. The correct answer is C.

   The remainder is the amount that is left over after you divide a problem into whole numbers. These whole numbers are referred to as factors. So, ask yourself what numbers can be calculated by multiplying by 3.

   1 × 3 = 3
   2 × 3 = 6
   3 × 3 = 9
   4 × 3 = 12

   12 is greater than 11, so the nearest product to 11 from the list above is 9.

   Finally, we subtract these two numbers to get the remainder.

   11 − 9 = 2

9. The correct answer is B.

   If there are 12 children and each one is supposed to receive 4 items, we can do the calculation as follows:

   12 children × 4 items per child = 48 items required in total

   Now subtract the total from the amount she already has in order to determine how many more she needs.

   48 items required in total − 40 items available = 8 items still needed

10. The correct answer is B.

    First, you need to determine the difference in degrees.

    90 degrees yesterday × 10% = 9 degrees cooler today

    Then subtract to get your answer.

    90 degrees yesterday − 9 degrees cooler today = 81 degrees today

11. The correct answer is C.

    First of all, you have to find out how many students were absent on Tuesday. To find the number of absent students, you have to multiply the total number of students in the class by the percentage of the absence for Tuesday.

    20 students in total × 5% = 1 student absent on Tuesday

    Now calculate the absences for Wednesday in the same way.

    20 students in total × 20% = 4 students absent on Wednesday

    The problem is asking you how many more students were absent on Wednesday than Tuesday, so you need to subtract the two figures that you have just calculated above.

    4 students absent on Wednesday − 1 student absent on Tuesday = 3 students

    So, 3 more students were absent on Wednesday.

12. The correct answer is D.

    Remember that when you see fractions that have different numbers on the bottom, you have to find the lowest common denominator (LCD).

In this problem, our LCD is 21.
So, we have to convert to fractions into the LCD. To calculate the LCD, you have to multiply the numerator and denominator by the same number.

$$\frac{1}{3} - \frac{1}{7} =$$

$$\left(\frac{1}{3} \times \frac{7}{7}\right) - \left(\frac{1}{7} \times \frac{3}{3}\right) =$$

$$\frac{7}{21} - \frac{3}{21} = \frac{4}{21}$$

13. The correct answer is B.
    The two tests are being given different percentages, so each test needs to have its own variable. A for test A and B for test B.
    Since A counts for 35% of the final grade, we set 35% to a decimal and put the decimal in front of the variable so that the variable will have the correct weight.
    So, the value of test A is .35A
    Test B counts for 65%, so the value of test B is .65B
    The final grade is the sum of the values for the two tests.
    So, we add the above products together to get our equation.
    .35A + .65B
14. The correct answer is C.
    Here is another question on setting up equations. The quantity of apples is represented by A and the quantity of bananas is represented by B.
    Apples are 20 cents each, while bananas cost 25 cents.
    So, the cost of apples is .20A
    The cost of bananas is .25B
    To get the total cost, we have to add the cost of the apples to the cost of the bananas
    .20A + .25B
15. The correct answer is A.
    Remember that for problems like this one, you need to get the terms that have a variable on one side of the equation. On the other side of the equation, you should have the whole numbers.
    STEP 1: Our first step is to deal with the whole numbers.
    $x - 3 + 5x = 33$
    $x - 3 + 3 + 5x = 33 + 3$
    $x + 5x = 36$
    STEP 2: Then simplify the numbers that have variables.
    $x + 5x = 36$
    $6x = 36$
    STEP 3: To solve the problem, remove the number in front of the variable by dividing.
    $6x \div 6 = 36 \div 6$
    $x = 6$
16. The correct answer is B.
    Remember that the perimeter is the measurement along the outside edges of the rectangle or other area.
    The formula for perimeter is as follows:
    P = 2W + 2L
    If the room is 12 feet by 10 feet, we need 12 feet × 2 feet to finish the long sides of the room and 10 feet × 2 feet to finish the shorter sides of the room.

(2 × 10) + (2 × 12) =
20 + 24 = 44

17. The correct answer is D.
Most months have 30 or 31 days. In this problem, we are being asked to do a calculation for a 2-month period, so we are dealing with 60 to 62 days.
For the purposes of estimation, we can use 60 days.
Ben uses a bag of dog food every 6 days.
So, we divide the total period by the number of days to get the required amount.
60 day period ÷ 6 days each bag = 10 bags needed for 60 days

18. The correct answer is B.
You will notice in this problem that we are dealing with two months, November and December.
STEP 1: Look at the total.
The total for the two months is 35 inches.
STEP 2: Determine whether one month is higher than the other.
In this problem, one month has 5 inches more than the other, so you have to subtract the difference first of all.
35 − 5 = 30
STEP 3: Now divide this amount by two to allocate each part to the two months.
30 ÷ 2 = 15
STEP 4: Look again at the problem to see if you are calculating the amount for the high month or the low month.
Here, the amount for November is lower than the amount for December.
So, we know that November had 15 inches of snowfall.
If the problem had asked for the higher month, you would then need to add back the difference.
So, December's snowfall is 15 + 5 = 20
STEP 5: Check your result by adding the amounts for the two months together.
15 + 20 = 35

19. The correct answer is B.
The problem is asking you for the amount that the average number of births per hospital in Johnson County for 2014 exceeded those for 1995.
STEP 1: First we have to calculate the average for 2014.
In order to calculate an average, you have to divide the total amount by the number of items in each data set.
For 2014, we have 240 total births and 15 hospitals in the data set.
240 ÷ 15 = 16 births on average per hospital for 2014
STEP 2: Now calculate the average for 1995
In our problem, this average is provided.
We can see that there were 12 births on average per hospital in Johnson County in 1995.
STEP 3: Now subtract the averages for the two years to get your answer.
16 − 12 = 4 more births per hospital in 2014

20. The correct answer is D.
This is a question on probability, which is a statistical measure.
STEP 1: Determine the total amount in the data set before any items are removed.
Here, we have a bag that contains 3 blue scarves, 1 red scarf, 4 green scarves, and 2 orange scarves.
3 + 1 + 4 + 2 = 10 items in the data set

STEP 2: Determine the numbers of items in the data set after items have been removed.
One scarf is removed.
10 – 1 = 9 items left in the data set
STEP 3: Determine the amount in the subset.
The problem is asking for the orange scarf subset. So, we have 2 orange scarves in the subset. Note that if the problem were asking you for the red scarf subset, you would have to subtract the item that has already been removed from the subset.
STEP 4: The probability is expressed as a fraction. The amount in the subset (2 orange scarves) goes on the top of the fraction and the amount of items left in the data set (9 items left) goes on the bottom.
So, the answer is $2/9$.

21. The correct answer is E.
This is another measurement problem.
STEP 1: You need to multiply the number of miles that she is going to travel by the amount of time it takes her to travel one mile.
17 minutes for 1 mile × 5 miles to travel = 85 minutes needed
STEP 2: Now express the result in hours an minutes, remembering of course that an hour has 60 minutes.
85 minutes – 60 minutes = 25 minutes left
So, the answer is 1 hour and 25 minutes.

22. The correct answer is B.
Remember to be careful with the negatives.
STEP 1: Deal with the negative numbers first.
−117 + (−25) + 45 =
−117 − 25 + 45 =
(−117 − 25) + 45 =
−142 + 45
STEP 2: Then deal with the positive number.
−142 + 45 =
−97

23. The correct answer is E.
STEP 1: Look at the chart to see which town is the highest.
Here, Eaton is the highest at 1621 feet above sea level.
STEP 2: Look at the chart to see which town is the lowest.
Appleton is the lowest at 687 feet below sea level.
STEP 3: Now add these two amounts together to find the total distance between the high point and the low point.
1621 + 687 = 2308
Note that if both points are above ground, you need to subtract the two amounts.

24. The correct answer is E.
The problem tells us that Clark County had 135,298 cases of infectious disease last year, while Davidson County had 207,121 cases.
STEP 1: Round each number up or down to the nearest thousand.
207,121 is rounded down to 207,000.
135,298 is rounded down to 135,000.
STEP 2: Subtract the two figures to estimate the difference.

207,000 − 135,000 = 72,000

25. The correct answer is C.
To get best estimate of 5,012 × 12, you need to round only the larger number up or down.
5,012 is rounded to 5,000
12 is not rounded in this case since the problem is asking for the *best* estimate.
Then multiply.
5,000 × 12 = 60,000

26. The correct answer is C.
This is a linear measurement problem.
STEP 1: Calculate the perimeter.
The field is 30 yards wide and 100 yards long.
The formula for perimeter is as follows:
P = 2W + 2L
(2 × 30) + (2 × 100) =
60 + 200 = 260 yards for the perimeter
STEP 2: Determine the linear total of all of the lines on the interior of the field. Be careful not to count the ends of the field again.
By counting the lines on the diagram, we can see that we have 9 lines on the interior of the field. Each line will be 30 yards in distance, since that is the width of the field.
9 lines × 30 yards each = 270 yards for the lines
STEP 3: Now add these two amounts together to get your answer.
260 + 270 = 530

27. The correct answer is E.
Remember to deal with the whole number first.
$3x - 4 - x = 12$
$3x (-4 + 4) - x = 12 + 4$
$3x - x = 16$
Then deal with the variable.
$3x - x = 16$
$2x = 16$
Then divide to get your answer.
$2x \div 2 = 16 \div 2$
$x = 8$

28. The correct answer is A.
We cannot calculate the number of students in the class with belts of particular colors because we do not know the color of the belts the new students.
The problem is telling us how many students there are in each group and how many of each group have left.
The problem also tells us how many students in total have joined, so we can calculate the new total number of students.

29. The correct answer is E.
Sam's final grade for a class is based on his scores from a midterm test (M), a project (P), and a final exam (F), but the midterm test counts twice as much as the project, and the final exam counts twice as much as the midterm. Therefore, we have to count variable M twice.
The value of the midterm is doubled and variable F is double of the midterm, so we have to count variable F 4 times.

So, the equation is: P + 2M + 4F

30. The correct answer is D.
The problem tells us that Bart rides at a rate of 12 miles per hour. We also know that he arrives in the town of Wilmington at 3:00 pm. The question is asking us how far Bart will be from Mount Pleasant at 5:00 pm.
STEP 1: Calculate the time difference.
5:00 pm – 3:00 pm = 2 hours difference
STEP 2: Calculate the distance traveled.
12 miles per hour × 2 hours = 24 miles traveled
STEP 3: Calculate the distance left.
The town of Mount Pleasant is 50 miles from Wilmington.
50 miles to travel – 24 miles traveled = 26 miles left

31. The correct answer is C.
The ticket office sold 360 more tickets on Friday than it did on Saturday. The office sold 2570 tickets in total during Friday and Saturday.
STEP 1: Subtract the excess.
2570 – 360 = 2210
STEP 2: Allocate the above figure to each day.
2210 ÷ 2 = 1105
STEP 3: Calculate Friday's amount by adding back in the excess.
1105 + 360 = 1465

32. The correct answer is D.
You need to find the relationships between the numbers provided in the chart in order to determine the missing value.
STEP 1: Consider whether a relationship between the numbers on the first row of the table can be found based on addition or subtraction.
Look at each of the sets of numbers on a line by line basis.
On the first line, we have 2 in the left column and 4 in the right column.
So, we can get to the value in the left column by adding 2.
STEP 2: Try out the value calculated in step 1 for the next row of numbers.
4 + 2 ≠ 16
STEP 3: If the relationship does not work for the second row of number we have to consider whether the relationship between the numbers is based on multiplication or division.
Returning to row 1 of the table, we can determine that: 2 × 2 = 4
STEP 4: Try this operation on the second row of numbers.
4 × 2 ≠ 16
STEP 5: Try to determine if any other relationship is possible.
Since 2 × 2 = 4 on the first row of the table, we can also try multiplying each subsequent number by itself.
STEP 6: Try this new relationship for the second and subsequent rows.
Row 2: 4 × 4 = 16
Row 4: 8 × 8 = 64
Row 5: 10 × 10 = 100
STEP 7: Calculate the value missing from row 3.
Row 3: 6 × 6 = 36

33. The correct answer is C.
    Tom was 5 feet tall at the beginning of the year, and his height increased by 10% this year.
    STEP 1: Calculate the beginning height in inches. Remember that there are 12 inches in a foot.
    5 feet × 12 inches per foot = 60 inches in height
    STEP 2: Calculate the increase in height.
    60 inches × 10% = 6 inches
    STEP 3: Calculate the new height by adding the increase to the number at the beginning.
    5 feet + 6 inches = 5 feet 6 inches
34. The correct answer is B.
    The problem is asking us what information is required in order to determine the time Mary arrived in Lordville.
    STEP 1: In order to determine the arrival time at a new destination, we need to know the time the person began the journey and the amount of time he or she traveled.
    From the problem, we know that Mary left home at 10:00 am.
    STEP 2: In order to know the amount of time a person travels, we need to know the amount of miles traveled and the speed.
    The problem tells us that she drove 150 miles to Newton at a constant rate of 70 miles per hour. However, her journey is ending in Lordville, so we also need to know the amount of miles traveled and the speed of traveling to Lordville.
    From the problem, we know that she drove 105 miles to Lordville after resting in Newton.
    STEP 3: Determine which information is missing.
    We do not know the speed she traveled to Lordville, so B is the correct answer.
35. The correct answer is E.
    This question is similar to the previous one.
    To calculate the average pace or speed, we need to know the speed for each journey.
    You will recall from the previous problem that in order to calculate the speed of travel, we need to know the distance traveled and the amount of time for the journey.
    The problem tells us that Dan rode his horse 2 miles to his neighbor's house and that it took 15 minutes for this journey.
    So, we have both the distance traveled and the amount of time for the first journey.
    The problem also states that Dan made a second journey, riding his horse 3 miles into town from his neighbor's house.
    So, we have the distance traveled for the second journey, but we do not have the amount of time for the second journey.
36. The correct answer is A.
    For addition problems like this one, remember that you can just add the numbers in front of the variables if both terms have the same variable.
    2H + 3H = 5H
37. The correct answer is E.
    If you have problems like this one, you should line all of the numbers up in a column to help you find the relationship. From the facts of the problem, we have 2,368,741 and 2,654,802. If you are unable to solve the problem visually, put the other numbers from the answer choices in between these two numbers to determine your answer.

Answer A:
2,368,741
2,281,414
2,654,802
2,281,414 is less than 2,368,741, so answer A is not the correct choice.

Answer B:
2,368,741
2,306,549
2,654,802

2,306,549 is less than 2,368,741, so answer B is not the correct choice.

Answer C:
2,368,741
2,367,988
2,654,802

2,367,988 is less than 2,368,741, so answer C is not the correct choice.

Answer D:
2,368,741
2,683,699
2,654,802

2,683,699 is greater than 2,654,802, so answer D is not the correct choice.

Answer E:
2,368,741
2,645,972
2,654,802

From the three figures above, we can see that answer E is correct since 2,645,972 is greater than 2,368,741 and less than 2,654,802.

38. The correct answer is C.
In order to calculate the percent of a number, we multiply the percent by the number.
Remember to convert the percent to a decimal by placing a decimal point two places from the right.
15% × 60 =
.15 × 60 =
60 × .15

39. The correct answer is C.
This question is similar to question 37 above.
Remember to place the numbers into column. For numbers that are decimals like in this problem, you can add zeroes to help you.
Answer A:
.00100
.00010
.00001

.00100 is greater than .00010, but the problem stipulates that .00100 is less than .00010, so answer A is not correct.

Answer B:
.00100
.00100
.00100

All of the figures provided in answer B are equivalent to one another, so it is not the correct answer.

Answer C:
.0001
.0010
.0100

Considering the three numbers above, .0001 is less than .0010, and .0010 is less than .0100, so answer C is correct.

Answer D:
.0010
.0010
.0100

Answer E:
.0100
.0100
.1000

The first two figures provided in answer D and in answer E are equivalent to one another, so they are not the correct answers.

40. The correct answer is B.

    If all of the numerators are 1, we know that the denominator of the unknown faction must lie between the denominators of the fractions stated in the problem.

    The problem asks: $1/3 > ? > 1/9$

    The denominator 3 from the equation above is greater than the denominator of 4 from answer B. The denominator of 4 from answer B is less than the denominator of 9 from the equation above. So, B is the correct answer.

    Note that if you are not able to find the correct answer from the options that have 1 in the numerator, you will need to calculate the lowest common denominator for all of the fractions.

41. The correct answer is D.

    The problem tells us that Carlos buys 2 pairs of jeans for $22.98 each, and then he decides to exchange both pairs of jeans for 3 sweaters which cost $15.50 each.

    STEP 1: Calculate the amount of money spent on the original purchase of the jeans.

    2 × $22.98 = $45.96

    STEP 2: Calculate the value of the items acquired in the exchange, which in this case, is the value of the sweaters.

    3 × $15.50 = $46.50

    STEP 3:

    Calculate the difference between the value of the items acquired and the amount of money originally spent.

    Value of the items acquired

    3 × $15.50 = $46.50

Amount of money originally spent
2 × $22.98 = $45.96
Difference:
(3 × $15.50) – (2 × $22.98)

42. The correct answer is A.
    Remember to put the numbers in columns if you are still not used to determining the value by visual inspection.
    Here, we see that .05 is less than .06, and that .050 is also greater than .006, so answer A is the correct choice.

43. The correct answer is C.
    If you look at the answer choices, you will see that they are given in the nearest increments of 5. So, we have to round the figures stated in the problem up or down to the nearest increment of 5.
    104 on Monday is rounded to 105.
    86 on Tuesday is rounded down to 85.
    81 is rounded down to 80.
    Then add these three figures together to get your result.
    105 + 85 + 80 = 270

44. The correct answer is D.
    We know that we have to round to the nearest hundredth.
    The hundredth decimal place is the number 2 positions to the right of the decimal.
    For example, .01 is 1 one hundredth.
    In our question, the first jump of 3.246 is rounded up to 3.25
    The second jump of 3.331 is rounded down to 3.33
    The third jump of 3.328 is rounded up to 3.33
    Then add these three figures together to get your answer.
    3.25 + 3.33 + 3.33 = 9.91

45. The correct answer is D.
    If there is practice today, we can conclude that it is a Tuesday or a Friday. The facts tell us that there will be no practice during the last full week of the month and that there will be no practice in the event of rain.

46. The correct answer is D.
    You have to find the relationship between the number given in each row in the left column and the corresponding number in the right column. "9:50 am to 10:36 am" represents a journey time of 46 minutes. 11:15 to 12:01 is also 46 minutes, and so on. If we go 46 minutes back from 5:51 pm, we arrive at 5:05 pm.

47. The correct answer is A.
    Look at the row of the table that displays information for the year 1965, which is the third row. We want to look at the last column of row 3 to find the figure for the entire world. The figure provided is 7413 tons.
    Next, we need to subtract the amount for the United Kingdom from the amount for the entire world in order to calculate the consumption outside the United Kingdom.
    The United Kingdom amount for 1965 was 3598 tons.
    7413 – 3598 = 3815 tons consumed outside the United Kingdom in 1965

48. The correct answer is C.
    For line graphs, look at the distance between the lines.

Here, we can see that the biggest difference, in other words, the largest gap between the two lines, is in Year 5.

Natural gas consumption was nearly 10,000 units in Year 5, while electricity consumption was approximately 6,000 units for the same year.

To calculate the difference, you need to subtract. 10,000 – 6,000 = 4,000

49. The correct answer is D.

    First, you need to add the stated percentages together to get the amount that is represented by the given slices of the pie chart.

    21% + 31 % + 7% + 14% + 10% = 83%

    Then subtract from 100% to get your answer.

    100% - 83% = 17% wheat production

50. The correct answer is D.

    For bar graphs, you will need to compare the height of the bars to one another.

    For Years 1 to 4, there is not much change from year to year.

    We can see from looking at the graph that the biggest growth was between years 4 and 5 or years 5 and 6.

    However, the difference between the bars for year 5 and year 6 is slightly less than the difference between the bars for year 4 and year 5.

    So, we have to choose answer D.

CBEST Practice Math Test 3

1. Captain Smith needs to purchase rope for his fleet of yachts. He owns 26 yachts and needs 6 feet 10 inches of rope for each one. How much rope does he need in total?
   A. 152 feet
   B. 177 feet 8 inches
   C. 257 feet 8 inches
   D. 260 feet
   E. 412 feet

2. Use the diagram below to answer the question that follows.

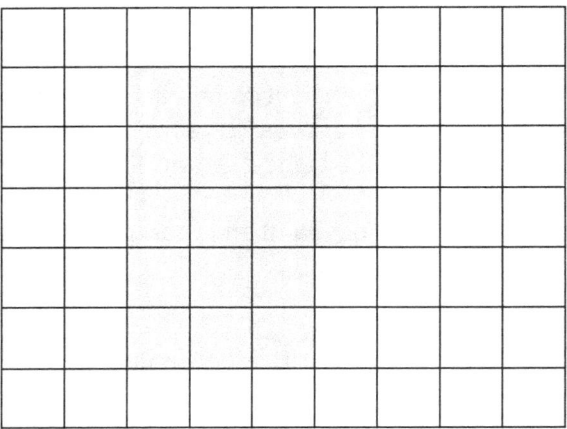

   Each square in the diagram above is one foot wide and one foot long. The gray area of the diagram represents the layout of New Town's water reservoir. What is the perimeter in feet of the reservoir?
   A. 16 feet
   B. 17 feet
   C. 18 feet
   D. 20 feet
   E. 32 feet

3. During this term, Tom got the following scores on his math tests: 98, 78, 89, 85, and 90. What is the average of Tom's scores?
   A. 78
   B. 85
   C. 88
   D. 89
   E. 98

4. What is the best unit of measure for expressing the weight of a bag of sugar?
   A. ounces
   B. inches
   C. pints
   D. quarts
   E. tons

5. Use the table below to answer the question that follows.

Part	Total Number of Questions	Number of Questions Answered Correctly
1	15	12
2	25	20
3	35	32
4	45	32

Chantelle took a test that had four parts. The total number of questions on each part is given in the table above, as is the number of questions Chantelle answered correctly. What was Chantelle's percentage of correct answers for the entire test?
A. 75%
B. 80%
C. 86%
D. 90%
E. 96%

6. Linda uses two bottles of ink every 5 days for her graphic design business. Approximately how many bottles of ink does she require for one month?
A. 2
B. 5
C. 6
D. 12
E. 75

7. Use the diagram below to answer the question that follows.

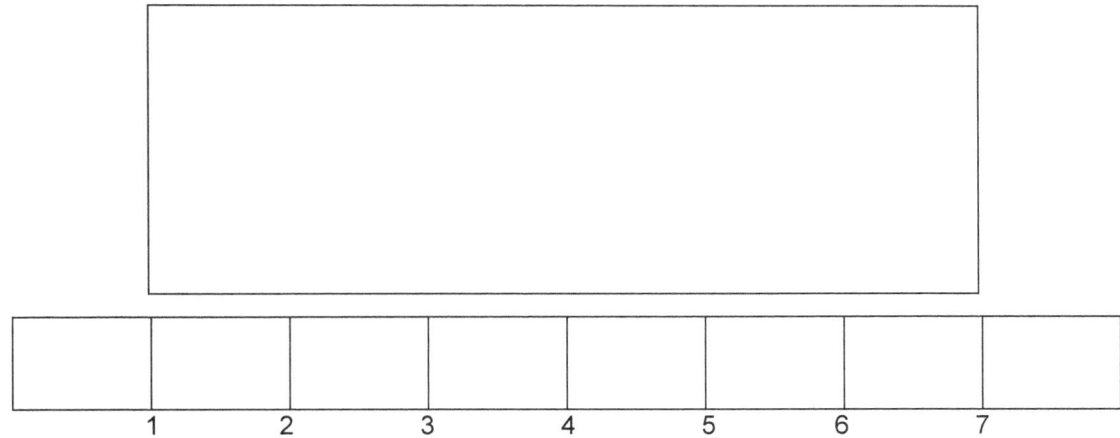

A scale drawing of a field is shown in the picture above. If the actual length of the field is 120 yards, what is the scale of the diagram?
A. 1 unit = 6 yards
B. 1 unit = 7 yards
C. 1 unit = 15 yards
D. 1 unit = 17 yards
E. 1 unit = 20 yards

8. A photograph measures 4 inches by 6 inches. Tom wants to make a wooden frame for the photo. He needs an extra inch of wood at each of the four corners in order to finish off the edges. What total length of wood will he need in order to complete the project?
   A. 10 inches
   B. 12 inches
   C. 16 inches
   D. 20 inches
   E. 24 inches

9. It takes Martha 4 hours and 10 minutes to knit one woolen cap. At this rate, how long will it take her to knit 12 caps?
   A. 40 hours
   B. 42 hours
   C. 46 hours
   D. 48 hours
   E. 50 hours

10. The Jones family needs to dig a new well. The well will be 525 feet deep, and it will be topped with a windmill which will be 95 feet in height. What is the distance from the deepest point of the well to the top of the windmill?
    A. 95 feet
    B. 430 feet
    C. 525 feet
    D. 610 feet
    E. 620 feet

11. Mr. Martin receives the following test report for a student who is in the tenth grade:

Raw Score	Percentile	Stanine	Grade Equivalent
68	54	6	9.2

    Which of the following statements provides a correct interpretation of the student's results?
    A. The student will be placed in the ninth grade.
    B. 68% of the other students taking the test scored better than this student.
    C. This student scored as well as or better than 32% of the other students taking the test.
    D. This student scored as well as or better than 54% of the other students taking the test.
    E. This student scored as well as or better than 46% of the other students taking the test.

12. At an elementary school, 3 out of ten students are taking an art class. If the school has 650 students in total, how many total students are taking an art class?
    A. 65
    B. 130
    C. 195
    D. 217
    E. 325

13. Mrs. Emerson plays a card game with the children in her class. She has 12 cards that have a picture of a fish, 15 cards that have a picture of a dog, 25 cards that have a picture of a cat, and 18 cards that have picture of a rabbit. She draws cards from the deck at random and shows them to the class. If the first card she draws is a rabbit, what is the probability that the next card will be a cat or a rabbit?
    A. $25/69$
    B. $25/70$
    C. $42/69$

D. 43/69
E. 43/70

14. What is the best estimate for 1,198 ÷ 29 ?
    A. 37
    B. 40
    C. 60
    D. 400
    E. 600

15. Simplify the following expression: −243 − (+ 225) + 13
    A. 5
    B. −5
    C. −31
    D. −455
    E. −481

16. What is the remainder for the following: 251 ÷ 13 ?
    A. 3
    B. 4
    C. 5
    D. 13
    E. 19

17. Mrs. Thompson is having a birthday party for her son. She is going to give balloons to the children. She has one bag that contains 13 balloons, another that contains 22 balloons, and a third that contains 25 balloons. If 12 children are going to attend the party including her son, and the total amount of balloons is to be divided equally among all of the children, how many balloons will each child receive?
    A. 3
    B. 4
    C. 5
    D. 6
    E. 7

18. Mount Pleasant is 15,238 feet high. Mount Glacier is 9,427 feet high. Which of the following is the best estimate of the difference between the altitudes of the two mountains?
    A. 5,700
    B. 5,800
    C. 5,900
    D. 6,000
    E. 6,100

19. A bookstore is offering a 15% discount on books. Janet's purchase would be $90 at the normal price. How much will she pay after the discount?
    A. $75.50
    B. $76.50
    C. $77.50
    D. $85.50
    E. $86.50

20. John is measuring plant growth as part of a botany experiment. Last week, his plant grew 7¾ inches, but this week his plant grew 10½ inches. What is the difference in growth in inches between the two weeks?
    A. 2¼ inches
    B. 2½ inches
    C. 2¾ inches
    D. 3¼ inches
    E. 3½ inches

21. At the beginning of a class, one-fourth of the students leave to attend band practice. Later, one half of the remaining students leave to go to PE. If there were 15 students remaining in the class at the end, how many students were in the class at the beginning?
    A. 30
    B. 40
    C. 45
    D. 50
    E. 80

22. Patty works 23 hours a week at a part time job for which she receives $7.50 an hour. She then gets a raise, after which she earns $184 per week. She continues to work 23 hours per week. How much did her hourly pay increase?
    A. 50 cents an hour
    B. 75 cents an hour
    C. $1.00 an hour
    D. $8.00 an hour
    E. $11.50 an hour

23. A packaging company places tape around a package that measures 4 inches in height, 5 inches in width, and 18 inches in length. One continuous piece of tape is placed around all four sides: on the top, bottom and both ends. Two further continuous pieces of tape are placed through the middle of the package around all four sides as shown in the illustration. How much tape is needed for this package?

    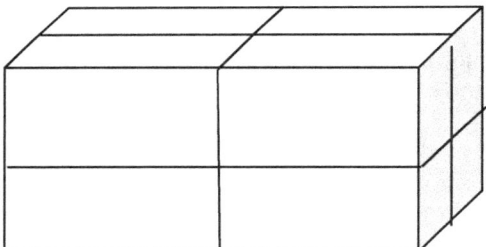

    A. 16 inches
    B. 20 inches
    C. 72 inches
    D. 100 inches
    E. 108 inches

24. Sheng Li is driving at 70 miles per hour. At 10:00 am, he sees this sign:

Washington	140 miles
Yorkville	105 miles
Zorster	210 miles

    He continues driving at the same speed. Where will Sheng Li be at 11:00 am?

A. 70 miles from Washington
B. 105 miles from Washington
C. 75 miles from Yorkville
D. 80 miles from Yorkville
E. 150 miles from Zorster

25. Mayumi spent the day counting cars for her job as a traffic controller. In the morning she counted 114 more cars than she did in the afternoon. If she counted 300 cars in total that day, how many cars did she count in the morning?
    A. 90
    B. 93
    C. 114
    D. 186
    E. 207

26. Solve for x: $3x + 5 - 2x = 15$
    A. 2
    B. 3
    C. 4
    D. 5
    E. 10

27. Shania is entering a talent competition which has three events. The third event (C) counts three times as much as the second event (B), and the second event counts twice as much as the first event (A). Which equation below can be used to calculate Shania's final score for the competition?
    A. A + 2B + C
    B. A + 2B + 3C
    C. A + 3B + 2C
    D. A + 2B + 6C
    E. A + B + C

28. Tiffany buys five pairs of socks for $2.50 each. The next day, she decides to exchange these five pairs of socks for four different pairs that cost $3 each. She uses this equation to calculate her refund: $(5 \times \$2.50) - (4 \times \$3)$
    Which equation below could she have used instead?
    A. $(5 \times 4) - (3 \times 2.50)$
    B. $\$2.50 - 4(\$3 - \$2.50)$
    C. $(5 \times 4) + (3 \times 2.50)$
    D. $\$3 - (4 \times \$2.50)$
    E. $\$3 - (5 \times \$2.50)$

29. Mr. Carlson needs to calculate 35% of 90.
    To do so, he uses the following equation:

    $$\frac{35 \times 90}{100}$$

    Which of the following could he also have used?
    A. $(35 \times 90) \div 100$
    B. $(35 \div 90) \times 100$
    C. $(35 - 90) \times 100$
    D. $90 \times .0035$
    E. $90 \div 35$

30. Read the information in the box below and answer the question that follows:

> - A health and beauty store has 90 bottles of shampoo for sale when the store opens for business on Monday morning.
> - These 90 bottles of shampoo consist of 15 bottles of strawberry-scented shampoo, 25 bottles of rose-scented shampoo, and 50 bottles of unscented shampoo.
> - At the close of business on Monday, 18 bottles of rose-scented shampoo remain in the store.

Which of the following facts can be determined from the information above?
A. The quantity of shampoo that the store normally offers for sale.
B. The average price of a bottle of shampoo.
C. The quantity of strawberry-scented shampoo sold on Monday.
D. The quantity of rose-scented shampoo sold on Monday.
E. The total quantity of shampoo left in the store at the close of business on Monday.

31. Which one of the following statements is correct?
A. $5/6 > 5/9 > 2/10$
B. $2/6 > 5/8 > 5/6$
C. $2/9 > 5/9 > 2/6$
D. $5/9 > 5/6 > 2/9$
E. $2/6 > 2/9 > 5/9$

32. Use the mathematical expression below to answer the question that follows: $1/6 < ? < 4/6$
Which of the following fractions would correctly complete the expression?
A. $1/3$
B. $1/9$
C. $2/3$
D. $6/9$
E. $8/12$

33. Read the problem below and then answer the question that follows.

> Tom and Mary are planning a cross-country trip. They plan to drive 300 miles each day for seven days. Their car can travel 25 miles on one gallon of gasoline. How much money in total will they need to pay for gasoline during their trip?

What piece of information is needed in order to answer the problem?
A. The amount of gasoline that the tank of the car can hold.
B. The total amount of miles that they will drive that week.
C. The price per gallon of gasoline.
D. The day of the week that their journey will begin.
E. The average speed of the car in miles per hour.

34. Read the problem below and then answer the question that follows.

> Paul leaves his house at 5:30 to go running. He runs 2 miles north through town, then continues 3 miles north out of town. He then runs south to his house along the same route. What is Paul's running pace?

What piece of information is needed in order to answer the problem?
A. The amount of steps that Paul makes.
B. The time that Paul returns home.
C. The length of Paul's stride.
D. The length of the return journey.
E. The total distance round-trip.

35. Use the chart below to answer the question that follows.

a	b
1.25	2.25
1.50	3.50
1.75	
2.00	6.00
2.25	7.25
2.50	8.50

The chart above demonstrates the relationship between variables *a* and *b*. What is the value of *b* that is missing from the chart?
A. 4.25
B. 4.50
C. 4.75
D. 5.00
E. 5.25

36. If the value of variable *x* is between 0.003 and 0.63, which one of the following could be variable *x*?
A. 0.0020
B. 0.0060
C. 0.6350
D. 0.7405
E. 0.0006

37. Which of the following mathematical expressions is equal to
$(x \times y) \div z$?
A. $(x \div y) \times z$
B. $(x \times z) \div y$
C. $(x \div z) \times y$
D. $(y \div x) \div z$
E. $z \times (x \div y)$

38. Which one of the following numbers is between 4,587,213 and 4,732,841?
A. 4,496,215
B. 4,567,633
C. 4,579,554
D. 4,587,125
E. 4,723,524

39. Use the information below to answer the question that follows.

- The supermarket is 12 miles away from the gas station.
- Tom's house is 18 miles away from the gas station.

Based on the information given above, which one of the following statements is correct?
A. Tom's house is 6 miles from the supermarket.
B. Tom's house is 12 miles from the supermarket.
C. Tom's house is no more than 18 miles from the supermarket.
D. Tom's house is exactly 18 miles from the supermarket.
E. Tom's house is no more than 30 miles from the supermarket.

40. Carl swam three races this week. The time of his first race was 36.21 seconds. The time of the second race was 35.78 seconds. The time of his third race was 34.93 seconds. If each of these times is rounded to the nearest one-tenth of a second, what is the estimate of Carl's total time for all three of the races?
    A. 106 seconds
    B. 106.8 seconds
    C. 106.9 seconds
    D. 107 seconds
    E. 107.1 seconds

41. Fatima drives 21 miles round trip every day between her home and her office. If her daily journey is rounded to the nearest 5 miles, which of the following is the best estimate of the total miles that Fatima drives in ten days?
    A. 150 miles
    B. 200 miles
    C. 210 miles
    D. 250 miles
    E. 300 miles

42. Use the information below to answer the question that follows.

    Classes will be held every Wednesday morning.
    If there are fewer than 3 children present for a class, the class will be canceled.
    If there is inclement weather, the class will be canceled.

    It is Wednesday morning and the class has been canceled. Which one of the following statements is correct?
    A. Fewer than three children were present for the class.
    B. There was inclement weather.
    C. Fewer than three children were present for the class and there was inclement weather.
    D. Fewer than three children were present for the class or there was inclement weather.
    E. More than three children were present for the class or there was inclement weather.

43. Use the information below to answer the question that follows.

    If the distance from his house to his destination is 5 miles or more, Jose uses his motorcycle.
    If the distance from his house to his destination is less than 5 miles but more than 1 mile, Jose uses his bicycle.
    If the distance from his house to his destination is 1 mile or less, Jose walks.

    Jose uses his bicycle to go to Manuel's house. Which one of the following statements could be true?
    A. Manuel's house is 1 mile or less from Jose's house.
    B. Manuel and Jose live 8 miles apart.
    C. Jose's house is at least 6 miles from Manuel's.
    D. Jose lives 4 miles from Manuel.
    E. The round trip between Jose and Manuel's houses is more than 10 miles.

44. Solve for x: $-12x + 15 + 16x = 31$
    A. 4
    B. 5
    C. 6
    D. 8
    E. 15

45. Use the information below to answer the question that follows.

> Classes in the morning last for 45 minutes, but classes in the afternoon last for 50 minutes.
> Lunch begins promptly at 12:30 pm and finishes promptly at 1:00 pm.
> There are 3 classes after lunch and 4 classes before lunch.
> There are no breaks between classes or between classes and lunch.

Which one of the following statements could be true?
A. Classes begin at 9:30am.
B. Classes begin at 10:00am.
C. The second class after lunch begins at 2:00pm.
D. The second class after lunch begins at 2:50pm.
E. The third class after lunch begins at 3:00pm.

46. Use the table below to answer the question that follows.

Waterloo Station Bus Timetable	
Departure Time	Arrival Time
9:18 am	11:06 am
10:32 am	12:20 pm
11:52 am	?
1:03 pm	2:51 pm

The bus journeys from Waterloo Station to a nearby town are always the same duration. What time is missing from the above timetable?
A. 12:40 pm
B. 1:34 pm
C. 1:40 pm
D. 1:48 pm
E. 1:51 pm

47. Use the chart below to answer the question that follows.

Diesel Oil Consumption in New York City and all of New York State in Tons		
Year	New York City	New York State
1989	1528	6547
1994	1782	6118
1999	1693	5974
2004	1521	6128
2009	1844	7029
2014	1732	8192

What was the diesel oil consumption in tons for all areas in the state outside of New York City for 1999?
A. 1693
B. 5974
C. 4281
D. 5019
E. 6547

48. Use the graph below to answer the question that follows.

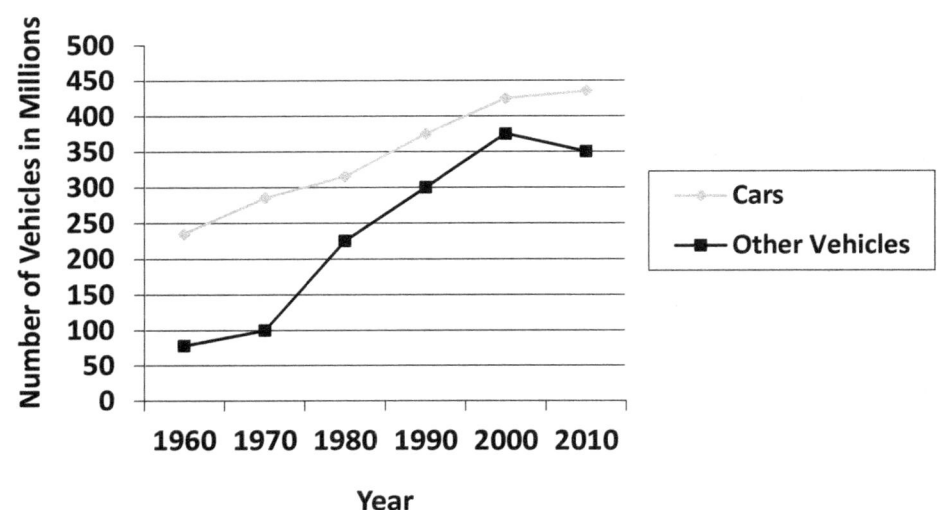

During what year was there the smallest difference between the numbers of cars owned and the number of other vehicles owned?
A. 1960
B. 1970
C. 1980
D. 1990
E. 2000

49. Use the chart below to answer the question that follows.

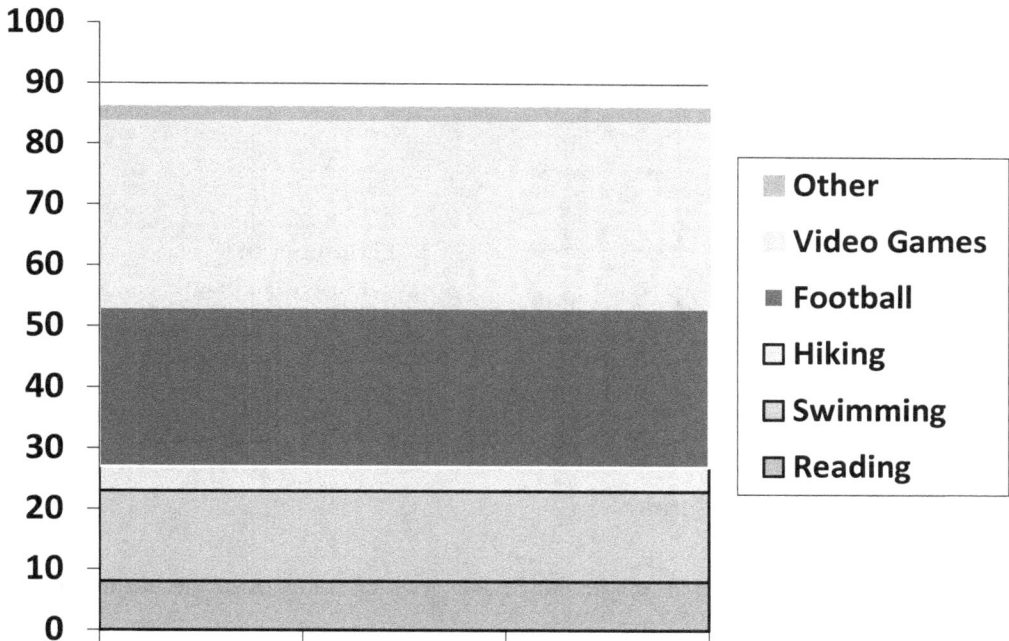

Favorite Student Hobbies

The percentage of students who enjoy participating in social media is omitted from the above chart. What percentage of students represented in the above chart have participating in social media as a favorite hobby?

A. 2%
B. 14%
C. 19%
D. 22%
E. 24%

50. Use the graph below to answer the question that follows.

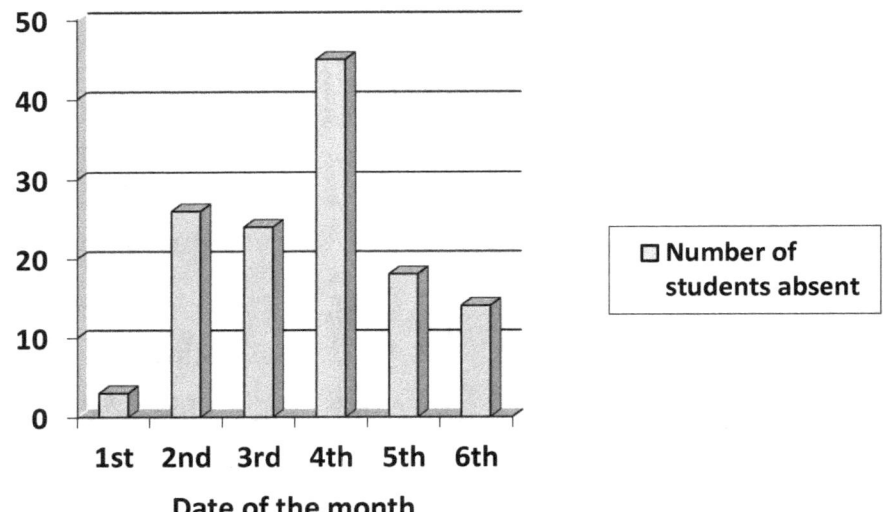

Student Absences at North High School

North High School experiences a flu epidemic. Between which dates does the number of student absences change the most?
A. the first to the second
B. the second to the third
C. the third to the fourth
D. the fourth to the fifth
E. the fifth to the sixth

CBEST Math Practice Test 3 – Answers

1. The correct answer is B.
   He owns 26 yachts and needs 6 feet 10 inches of rope for each one.
   Convert the feet and inches measurement to inches.
   6 feet 10 inches =
   (6 × 12) + 10 inches =
   72 + 10 = 82 inches
   Then multiply by the number of items.
   26 × 82 = 2132 inches of rope needed
   Then convert back to feet and inches.
   2132 inches ÷ 12 = 177 feet 8 inches

2. The correct answer is C.
   Count how many blocks lie along the outer edges of the shaded area in order to get your result.
   Top boundary = 4 feet
   Left side boundary = 5 feet
   Bottom boundary = 3 feet
   Right boundary = 6 feet (Don't forget to count the piece shaped like the upside-down "L" on the right.)
   Then add these amounts to get your result.
   4 + 5 + 3 + 6 = 18 feet

3. The correct answer is C.
   To get the average, add up all of the items.
   98 + 78 + 89 + 85 + 90 = 440
   There are five scores, so there were five tests.
   Divide the total points by the number of tests in order to get the average.
   440 ÷ 5 = 88

4. The correct answer is A.
   Sugar is a dry item, so it is measured by weight. Remember that wet items are measured in pints and quarts, while dry items are measured in ounces and pounds, or tons in the case of extremely large quantities.
   Feet and inches are linear measurements; they are not used for weight.
   A bag is sugar is a small item, so the correct answer is ounces.

5. The correct answer is B.
   First of all, add up the number of questions answered correctly.
   12 + 20 + 32 + 32 = 96
   Then add up the total number of questions.
   15 + 25 + 35 + 45 = 120
   Now divide the number of questions answered correctly by the total number of questions to get her percentage score.
   96 ÷ 120 = 80%

6. The correct answer is D.
   Assuming there are 30 days in the month, we can divide as shown.
   30 days ÷ 5 days per 2 bottles =
   6 days × 2 bottles = 12 bottles needed for 1 month

7. The correct answer is E.
   STEP 1: Determine the length of the field in units.
   We can see that the right-hand side of the field is on the number 7. However, the field is not positioned over the first unit of the ruler.
   So, the field is 6 units long.
   STEP 2: Divide the actual length of the field by the units.
   120 ÷ 6 = 20
   STEP 3: Express the answer in units and yards.
   1 unit = 20 yards
8. The correct answer is E.
   Measure the length along the top and bottom of the frame, as well as the length of both sides in order to get the basic perimeter.
   Top = 4 inches
   Bottom = 4 inches
   Left side = 6 inches
   Right side = 6 inches
   Total perimeter: 4 + 4 + 6 + 6 = 20 inches
   Now add in the 4 extra inches for the four corners.
   20 + 4 = 24 inches
9. The correct answer is E.
   STEP 1: Convert into minutes the amount of time required to make one cap.
   4 hours and 10 minutes =
   (4 × 60) + 10 =
   240 + 10 = 250 minutes needed to make one cap
   STEP 2: Multiply by the total output.
   250 minutes × 12 caps = 3000 minutes
   STEP 3: Convert the total amount of minutes back to hours and minutes
   3000 minutes ÷ 60 = 50 hours
10. The correct answer is E.
    Add the feet above ground to the feet below ground to get the total distance.
    525 + 95 = 620 feet
11. The correct answer is D.
    Remember the following basic aspects of standardized score reports. Raw score is the number of questions answered correctly, assuming that each correct response receives one point.
    Percentile means that the student scored as well as or better than this percentage of other students taking the test.
    Stanine is short for "Standard nine." Stanine shows the student's score on a scale of 1 to 9.
    The grade equivalent shows the grade-level of school the student's performance. For example, a grade equivalent of 9.2 means that the student performed on the exam at the level of a 9th grade student. In our case, we cannot say with certainty that this student will be placed in ninth grade classes since this depends on the school's policies. The report states that the student's percentile score is 54. Therefore, the student scored as well as or better than 54% of the other students taking the test.
12. The correct answer is C.
    Three out of ten students are taking the class. So, here we have the proportion 3 to 10.

STEP 1: Divide the total number of students by the second number in the proportion to get the number of groups.
650 ÷ 10 = 65 groups
STEP 2: Multiply the number of groups by the first number in the proportion in order to get the result.
3 × 65 = 195 art students

13. The correct answer is C.
Here is another probability question.
STEP 1: Remember that your first step is to determine the number of items in the data set, before any items are removed.
Mrs. Emerson has 12 cards that have a picture of a fish, 15 cards that have a picture of a dog, 25 cards that have a picture of a cat, and 18 cards that have picture of a rabbit.
12 + 15 + 25 + 18 = 70
STEP 2: Determine the amount in the data set after any items have been removed.
We know that the first card she draws is a rabbit, so she has taken one item from the data set.
70 − 1 = 69
STEP 3: Determine the amount in the subset.
Our subset is cards with cats or rabbits. Before any items were removed, we had 25 cards with picture of a cat, and 18 cards with picture of a rabbit. Then one card with a rabbit was removed. So, add and subtract as shown.
25 + 18 − 1 = 42 cat or rabbit cards remaining
STEP 4: The probability is expressed as a fraction, with the subset on the top and the data set on the bottom.
$^{42}/_{69}$

14. The correct answer is B.
We can see from the answer choices that we are rounding to the nearest increment of 10.
1,198 is rounded up to 1,200
29 is rounded up to 30
Now do the operation.
1,200 ÷ 30 = 40

15. The correct answer is D.
Remember to be careful with the negatives.
−243 − (+ 225) + 13 =
−243 − 225 + 13 =
−468 + 13 =
−455

16. The correct answer is B.
251 ÷ 13 = ?
Do long division as shown in order to get the remainder.

```
      19
13)251
    13
    121
    117
      4
```

17. The correct answer is C.
    STEP 1: Add the items together to get the total amount of items available.
    13 + 22 + 25 = 60 balloons in total
    STEP 2: Divide the amount of items available by the number of people.
    60 ÷ 12 = 5
18. The correct answer is B.
    We have to subtract to find the difference in height between the two mountains.
    First of all, round each number up or down.
    Looking at the answer choices, we can see that we need to round to the nearest increment of 100.
    15,238 is rounded down to 15,200
    9,427 is rounded down to 9,400
    Now subtract to get your answer.
    15,200 − 9,400 = 5,800
19. The correct answer is B.
    STEP 1: Determine the value of the discount by multiplying the normal price by the percentage discount.
    $90 × 15% = $13.50 discount
    STEP 2: Subtract the value of the discount from the normal price to get the new price.
    $90 − $13.50 = $76.50
20. The correct answer is C.
    STEP 1: You can express the fractions as decimals for the sake of simplicity.
    10½ = 10.50
    7¾ = 7.75
    STEP 2: Then subtract to find the increase.
    10.50 − 7.75 = 2.75
    STEP 3: Then convert back to a mixed number.
    2.75 = 2¾
21. The correct answer is B.
    One-fourth of the students leave to attend band practice. Later, one half of the remaining students leave to go to PE.
    We know we have 15 students remaining in class after the others have left.
    So, work backwards based on the facts given.
    STEP 1: We have 15 students just after half of the students have left to go to PE, so divide 15 by one-half. So, if x represents the number of remaining students after the group of students have left for PE, we have the following equation.
    15 = x − (x × ½)
    15 = x − ½ x
    15 = ½ x
    15 × 2 = ½ x × 2
    30 = x
    So, there were 30 students in class before PE class
    STEP 2: We have 30 students after ¼ of them have gone to band, so we have the following equation.
    30 = x − (x × ¼)
    30 = x − ¼ x

$30 = ¾ x$
$30 × 4 = ¾ x × 4$
$120 = 3x$
$120 ÷ 3 = 3x ÷ 3$
$40 = x$
So, there were 40 students in the class at the beginning.

22. The correct answer is A.
    After her raise, she earns $184 per week. She continues to work 23 hours per week.
    STEP 1: Determine the new hourly rate.
    $184 ÷ 23 hours = $8 per hour
    STEP 2: Determine the change in the hourly rate.
    $8 - $7.50 = 50 cents per hour

23. The correct answer is E.
    STEP 1: We are covering all four sides, so we need to multiply each of the dimensions by 4.
    18 × 4 = 72
    5 × 4 = 20
    4 × 4 = 16
    STEP 2: Add all of the above figures together to get your result.
    72 + 20 + 16 = 108

24. The correct answer is A.
    STEP 1: Determine the distance traveled.
    If he is traveling 70 miles an hour, he will have traveled 70 miles after one hour has passed.
    STEP 2: Determine the distance from the towns listed on the sign, considering that he has traveled for one hour.
    Washington: 140 – 70 = 70 miles from Washington
    Yorkville: 105 – 70 = 35 miles from Yorkville
    Zorster: 210 – 70 = 140 miles from Zorster
    STEP 3: Compare the above figures to your answer choices to get your result. After an hour, he is 70 miles from Washington, so A is correct.

25. The correct answer is E.
    STEP 1: Subtract the excess from the total.
    300 – 114 = 186
    STEP 2: Allocate the difference into its respective parts.
    We are dividing the day into two parts: morning and afternoon. There were 186 cars in total without the excess, so divide this into two parts.
    186 ÷ 2 = 93
    STEP 3: Determine the amount for the larger part.
    There were 114 more cars in the morning, so add this back in.
    93 + 114 = 207 cars in the morning

26. The correct answer is E.
    Remember to deal with the integers, and then deal with the variable.
    $3x + 5 – 2x = 15$
    $3x + 5 – 5 – 2x = 15 – 5$

    $3x – 2x = 10$
    $x = 10$

27. The correct answer is D.
    Work out the equation based on the facts provided in the problem.
    The second event (B) counts twice as much as the first event (A), so we need to represent the value of the second event as 2B.
    The third event (C) counts three times as much as the second event, so we need to multiply the value of the second event by 3.
    2 × 3 = 6
    So, the value of the third event is 6C.
    Therefore, the equation is A + 2B + 6C.
28. The correct answer is B.
    Here is another exchange problem.
    STEP 1: Think about the value of the four pairs of socks she is getting in the exchange. These socks cost 50 cents more each than the pairs she has already bought. So, we can express the difference in value of those four pairs of socks as: 4 × ($3 - $2.50)
    STEP 2: Take into account the value of the extra pair of socks. She paid $2.50 for a fifth pair of socks, but she is only getting four pairs back on the exchange, so she is owed money back for that part of the purchase.
    Therefore, we can calculate the refund owing as $2.50 − 4($3 - $2.50)
29. The correct answer is A.
    The line in any fraction can be treated as the division symbol. Accordingly, we can divide by the denominator, which is 100 in this case.
    $$\frac{35 \times 90}{100} = (35 \times 90) \div 100$$
30. The correct answer is D.
    We don't know how many bottles of strawberry or unscented shampoo were sold. Nor do we know what the store sells normally. So, we cannot calculate the total quantity of shampoo left unsold in the store when it closes on Monday. We can only calculate the quantity of rose-scented shampoo sold since the facts tell us how many bottles of rose-scented shampoo remain in the store at the close of business.
31. The correct answer is A.
    Rule each answer choice out by the process of elimination.
    Remember these shortcuts when dealing with fractions:
    Fractions with the same numerators:
    If the numerators on the tops of two fractions are the same, the fraction with the smaller denominator is actually the greater fraction.
    For example: ½ > ¼
    If any of the inequalities in the answer choices have the same numbers in their numerators, you can then just compare the denominators in the answer choices in order to determine which fraction is greater.
    This is the case with answer choices B and D, so let's evaluate them.
    $2/6 > 5/8 > 5/6$ – Answer B is incorrect because $5/8$ is less than $5/6$.
    $5/9 > 5/6 > 2/9$ – Answer D is incorrect because $5/9$ is less than $5/6$.
    Fractions with the same denominators:
    On the other hand, if the denominators on the bottoms of two fractions are the same, the fraction with the larger numerator is the greater fraction.
    This is the case with answer choices C and E, so let's look at them next.

$^2/_9 > {^5/_9} > {^2/_6}$ – Answer C is incorrect because $^2/_9$ is less than $^5/_9$.

$^2/_6 > {^2/_9} > {^5/_9}$ – Answer E is also incorrect for the same reason.

Now have a look at the other answer choices.

Answer choice A is as follows: $^5/_6 > {^5/_9} > {^2/_{10}}$

Using the principles above, we can see that $^5/_6 > {^5/_9}$ from answer A is correct.

So, next you can evaluate whether $^5/_9 > {^2/_{10}}$ from answer A is also correct.

First, we have to find the lowest common denominator.

$$\frac{5}{9} > \frac{2}{10}$$

$$\left(\frac{5}{9} \times \frac{10}{10}\right) > \left(\frac{2}{10} \times \frac{9}{9}\right)$$

$$\frac{50}{90} > \frac{18}{90}$$

Both parts of the inequality are correct, so A is the correct answer.

32. The correct answer is A.

    Check each answer option one by one.

    From answer A, we can see that $^1/_6$ is less than $^1/_3$.

    Then put the other part of the inequality into the LCD to check the answer.

    $$\frac{1}{3} < \frac{4}{6}$$

    $$\left(\frac{1}{3} \times \frac{2}{2}\right) < \frac{4}{6}$$

    $$\frac{2}{6} < \frac{4}{6}$$

    Both parts of the inequality are correct, so answer A is the correct answer.

33. The correct answer is C.

    In order to solve this problem, we would need to multiply the number of gallons of gasoline used per day by the cost of gasoline per gallon by the number of days traveled in order to calculate the total cost.

    From these required facts, we are lacking the price of gasoline per gallon.

34. The correct answer is B.

    We know that Paul will have run ten miles when he finishes since he runs 5 miles north, then returns and goes 5 miles south.

    The question is asking about his running pace or speed.

    In order to know speed, we need to know the distance traveled and the amount of time it takes to travel the distance.

    So, we know the distance, but not the time.

    Accordingly, we would need to know what time he gets back home in order to solve the problem.

35. The correct answer is C.

    Remember to check the relationship between the numbers in each column on a row-by-row basis.

Here, we can solve by addition. The pattern is that the row number is added to the value in column a in order to find its value in column b.
So, for row 1: 1.25 + 1 = 2.25
For row 2: 1.50 + 2 = 3.50
For row 3: 1.75 + 3 = 4.75

36. The correct answer is B.
Remember to line all of the numbers up in a column if you have difficulties solving problems like this visually.
By visual inspection, we can see that answers A and E are too small, while answers C and D are too large.
0.003 < 0.006 < 0.63, so answer B is correct.

37. The correct answer is C.
Remember that the division symbol is the same calculation mathematically as the line in a fraction.
So, express the equations as fractions to check the answer.

$$(x \times y) \div z = \frac{x \times y}{z} = \frac{xy}{z}$$

From answer C, we can determine that:

$$(x \div z) \times y = \frac{x}{z} \times y = \frac{xy}{z}$$

38. The correct answer is E.
Answers A to D are too small. Putting the numbers in a column, we can check that answer E is correct:
4,587,213
4,723,524
4,732,841

39. The correct answer is E.
For questions like this, you will recall that the points could lie on one continuous strait path like a line. Alternatively, the points could be laid out more like a triangle. However, the distance between points will always be greater when the points are linear.
If the points are linear, then the maximum distance will be calculated as follows:
12 miles + 18 miles = 30 miles

40. The correct answer is C.
Remember that the tenth is the decimal just to the right of the decimal point.
So, we need to round as required.
The first race was 36.21 seconds, which is rounded down to 36.2
The second race was 35.78 seconds, which is rounded up to 35.8
The third race was 34.93 seconds, which is rounded down to 34.9
Now add these figures together.
36.2 + 35.8 + 34.9 = 106.9

41. The correct answer is B.
We round the daily distance to 20, and then multiply by 10 to get the estimate of 200.

42. The correct answer is D.
The second fact tells us that if there are fewer than 3 children present for a class, the class will be canceled.

The third fact tells us that if there is inclement weather, the class will also be canceled.

43. The correct answer is D.
The second fact tells us that if the distance from his house to his destination is less than 5 miles but more than 1 mile, Jose uses his bicycle. If Jose uses his bicycle to go to Manuel's house, then it might be possible that Jose lives 4 miles from Manuel.

44. The correct answer is A.
$-12x + 15 + 16x = 31$
$-12x + 15 - 15 + 16x = 31 - 15$
$-12x + 16x = 16$
$4x = 16$
$4x \div 4 = 16 \div 4$
$x = 4$

45. The correct answer is A.
If classes last for 45 minutes and there are 4 classes before lunch, the morning classes last for 3 hours in total.
If lunch is at 12:30, it is therefore possible for classes to begin at 9:30.

46. The correct answer is C.
Each journey is 108 minutes (1 hour and 48 minutes) in duration.
So, we need to add 108 minutes to the departure time of 11:52 to get the arrival time of 1:40.

47. The correct answer is C.
Subtract the New York City amount from the total for the entire state for 1999 to get the answer.
$5974 - 1693 = 4281$

48. The correct answer is E.
Look to see which year has the smallest gap between the two lines.

49. The correct answer is B.
If the entire graph were complete, we would have 100%.
At the moment, we have 86%, so we need to subtract.
$100\% - 86\% = 14\%$

50. The correct answer is D.
Notice that the question in asking about the biggest change in general, rather than the largest increase or the largest decrease. Visually, we can see that the gap between months 4 and 5 is the greatest.

CBEST Practice Math Test 4

1. Which of the following shows the numbers ordered from least to greatest?
   A. −1/4 , 1/8 , 1/6 , 1
   B. −1/4 , 1/8 , 1 , 1/6
   C. −1/4 , 1/6 , 1/8 , 1
   D. −1/4 , 1 , 1/8 , 1/6
   E. 1 , 1/6 , 1/8 , −1/4

2. When 1523.48 is divided by 100, which digit of the resulting number is in the tenths place?
   A. 1
   B. 2
   C. 3
   D. 4
   E. 5

3. You work as a climatologist and need to calculate the average high temperature in one city over a five-day period in degrees Celsius. However, the high temperatures are reported in Fahrenheit. You have collected the following data: Day 1: 72° F; Day 2: 68° F; Day 3: 65° F; Day 4: 82° F; Day 5: 81° F. What was the approximate average high temperature in degrees Celsius? °C = 0.56(°F − 32)
   A. 74°C
   B. 73°C
   C. 41°C
   D. 32°C
   E. 23°C

4. An art and craft store received $7,375 for sales of a certain type of scrapbook this year. If these scrapbooks were sold for $59 each, how many of them were sold this year?
   A. 135
   B. 125
   C. 120
   D. 75
   E. 65

5. One private airplane flew at a constant speed, traveling 780 miles in 2 hours. How many miles did this plane travel in the last 40 minutes of its journey?
   A. 120
   B. 180
   C. 200
   D. 260
   E. 240

6. A horse ran 12 furlongs in 2 minutes and 48 seconds. Assuming that the same amount of time was spent on each furlong, how many seconds does it take the horse to run one furlong?
   A. 0.014 seconds
   B. 0.14 seconds
   C. 1.40 seconds
   D. 14 seconds
   E. 140 seconds

7. A report shows that 2 out of every 20 employees of a particular company are interested in applying for a promotion. If the company has 480 employees in total, how many employees are interested in applying for a promotion?
   A. 20
   B. 24
   C. 42
   D. 48
   E. 84

8. An item costs $22 each if the customer collects it in person from the store, and an extra $3 for postage and handling is charged per item if the customer wants the item sent by courier. This week, 32 customers purchased this item and requested that the item be sent by courier. How much money in total did the store collect on the items sold to these 32 customers?
   A. $800
   B. $704
   C. $575
   D. $525
   E. $96

9. $6/25$ of the inventory has been sold this month. Approximately what percentage of the inventory has been sold?
   A. 0.24%
   B. 2.40%
   C. 24.0%
   D. 4.167%
   E. 6.667%

10. A vat contains 163.75 units of red colorant, 107.50 units of blue colorant, 91.25 units of yellow colorant, and 10.30 units of black colorant. Which of the following represents, in terms of units, how full the vat is after these 4 colorants have been placed in it?
    A. 362.50
    B. 371.50
    C. 372.80
    D. 373.50
    E. 374.50

11. A customer who owns a small hotel has ordered 10 new quilts. Each quilt requires 2 yards of red fabric for the front, 1 yard of blue fabric for the front, and a further 3 yards of blue fabric for the back. The quilts need to have an embellishment in gold, and a total amount of 6 yards of gold fabric is needed to make the embellishments for all 5 quilts. Each quilt also has edging in white, and half a yard of white material is needed for the edging for each quilt. How many yards of fabric in total will be needed to complete this order?
    A. 7.7
    B. 77
    C. 3.85
    D. 38.50
    E. 3.95

12. A store sells domestic cleaning products. A certain type of liquid cleaner is sold in increments of 1/4 of a cup. Each 1/4 of a cup costs 50 cents. One customer buys $10\frac{1}{4}$ cups of this cleaner. How much will she pay for this purchase?
    A. $5.13

B. $5.50
C. $10.50
D. $20.50
E. 21.50

13. The cost of sales figures each month for a company's first five months of business this year were: $723, $618, $576, $812, and $984. What is the best estimate of the total cost of sales for the first five months of business this year?
A. $2,600
B. $2,700
C. $3,600
D. $3,700
E. $4,700

14. A liquid ingredient is stored in 5-quart containers. There are two partially-full containers, one with $4^{3}/_{8}$ quarts and another with $3^{7}/_{8}$ quarts. How many quarts are there in total in these two containers?
A. $1^{1}/_{4}$
B. 7
C. $7^{1}/_{8}$
D. $8^{1}/_{4}$
E. 9

15. A small factory uses tarpaulin to make covers for farm implements. There was $12^{7}/_{16}$ yards of tarpaulin at the start of the day. At the end of the day, $8^{9}/_{16}$ yards of tarpaulin is left. Which amount below represents the amount of tarpaulin used this day in yards?
A. $2^{14}/_{16}$
B. $3^{1}/_{8}$
C. $3^{7}/_{8}$
D. $4^{7}/_{8}$
E. 5

16. Abdul purchased 80 items for sale, and he has sold 0.75 of them in relation to the total purchased. How many items does he have left after making these sales?
A. 10 items
B. 20 items
C. 25 items

D. 40 items
E. 42 items

17. A footwear store can purchase 325 pairs of tennis shoes from its normal supplier for $4 a pair. It can get the same 325 pairs of shoes from a second supplier for $1,250 plus 6% sales tax, or from a third supplier for $1,290. How much will the store pay to get the best deal?
A. $1,250.00
B. $1,290.00
C. $1,300.00
D. $1,367.40
E. $1,368.40

18. 4 out of every 5 employee-satisfaction questionnaires have been completed and returned. If a company has 250 total employees, and every employee must complete and return the questionnaire, how many questionnaires have not been completed and returned?
    A. 4
    B. 5
    C. 50
    D. 200
    E. 210

19. A flower store sells poinsettia plants for $20 during December and for $12 during January. In December, 55 customers purchased poinsettias, and 20 customers purchased them in January. How much money did the store receive for poinsettia sales during December and January?
    A. $240
    B. $1,060
    C. $1,100
    D. $1,340
    E. $1,360

20. In the last step of doing a calculation, Wei Li added 92 instead of subtracting 92. What shortcut can Wei Li perform in order to get the correct calculation?
    A. Subtract 46 from his erroneous result.
    B. Add 92 to his erroneous result.
    C. Subtract 92 from his erroneous result.
    D. Add 184 to his erroneous result.
    E. Subtract 184 from his erroneous result.

21. A caterpillar travels 10.5 inches in 45 seconds. How far will it travel in 6 minutes?
    A. 45 inches
    B. 63 inches
    C. 64 inches
    D. 84 inches
    E. 90 inches

22. The ratio of males to females in the senior year class of Carson Heights High School was 6 to 7. If the total number of students in the class is 117, how many males are in the class?
    A. 48
    B. 54
    C. 56
    D. 58
    E. 63

23. Results from a questionnaire administered to customers of a particular supermarket show that 4 out of 7 customers prefer toffee-flavored ice cream to coffee-flavored ice cream. Based on these results, if 1,540 customers purchased one of these two flavors of ice cream, how many of them would have purchased coffee-flavored ice cream?
    A. 220
    B. 420
    C. 560
    D. 660
    E. 880

24. An employment agency for temporary employees charges clients $15 per hour for each hour the temporary employee works. The agency pays each temporary employee $12 an hour and retains the difference as a commission. The agency had 10 employees who worked 40 hours each this week. How much did the agency make on commission for these 10 employees this week?
A. $30.00
B. $120.00
C. $1,200.00
D. $4,800.00
E. $5,200.00

25. 49 out of the 50 items in a company's product line had above average sales this month. What percentage of the items in the product line had above average sales this month?
A. 0.098%
B. 0.98%
C. 9.80%
D. 98%
E. 980%

26. Sales each day for the past five days have been as follows: $90, $85, $85, $105, $110. What was the daily average sales amount during this five-day period?
A. $25
B. $85
C. $90
D. $95
E. $105

27. A fabric store sells ribbon in 3-inch or one-foot increments. One customer wanted two types of ribbon, and she bought $8^{3}/_{4}$ feet of one type of ribbon and $7^{1}/_{2}$ feet of another type. How much ribbon did this customer buy in total?
A. 7 feet and 6 inches
B. 8 feet and 9 inches
C. 15 feet and 3 inches
D. 16 feet and 3 inches
E. 17 feet and 3 inches

28. Hours spent on a work order are recorded by the tenth of an hour in 6-minute increments. For a particular work order, $28^{3}/_{10}$ hours in total have been budgeted. $7^{9}/_{10}$ hours have already been spent on the work order. Which amount below represents the amount of time left for this work order?
A. $36^{1}/_{5}$
B. $35^{6}/_{10}$
C. $20^{2}/_{5}$
D. $21^{3}/_{5}$
E. $22^{3}/_{5}$

29. A measurement of 116 feet is how many inches longer than a measurement of 36 yards?
A. 8
B. 80
C. 96
D. 960
E. 3,744

30. The ages of 5 siblings are: 2, 5, 7, 12, and x. If the average age of the 5 siblings is 8 years old, what is the age (x) of the 5th sibling?
    A. 8
    B. 10
    C. 12
    D. 14
    E. 16

31. A recipe of the ingredients needed to make 4 brownies is provided below.

    > Brownie recipe
    > ¼ cup of flour
    > ½ cup of sugar
    > ¼ cup of butter
    > 3 tablespoons of cocoa powder
    > ¼ teaspoon of baking powder
    > ½ teaspoon of vanilla extract

    How much cocoa powder and baking powder together is needed to make 12 brownies? (1 tablespoon = 3 teaspoons)
    A. 9¼ teaspoons
    B. 27¼ teaspoons
    C. 27½ teaspoons
    D. 27¾ teaspoons
    E. 28¾ teaspoons

32. Look at the table below and then answer the question that follows.

	Live in the City	Live in the Country
Factory workers	35	23
Office workers	42	20

    The table above shows data on the distribution of residents of a particular county, according to their types of work and the locations of their residences. If one of the residents from this sample is selected at random, what is the probability that the resident will be an office worker who lives in the country?
    A. $1/7$
    B. $1/6$
    C. $1/3$
    D. $23/120$
    E. $42/120$

33. An illusionist pulls colored scarves out of a hat at random. The hat contains 5 red scarves and 6 blue scarves. The other scarves in the hat are green. If a scarf is pulled out of the hat at random, the probability that the scarf is red is $1/3$. How many green scarves are in the hat?
    A. 3
    B. 4
    C. 5
    D. 6
    E. 7

34. $\frac{3}{4}x - 2 = 4$, $x = ?$
    A. $\frac{8}{3}$
    B. $\frac{1}{8}$
    C. 8
    D. −8
    E. 24

35. Which of the following steps will solve the equation for x: 4x – 3 = 2
    A. Add 3 to both sides of the equation, and then divide both sides by 4.
    B. Add 3 to both sides of the equation, and then subtract 4 from both sides.
    C. Add 2 to both sides of the equation, and then divide both sides by 4.
    D. Subtract 2 from both sides of the equation, and then divide both sides by −3.
    E. Divide both sides of the equations by 4, and then subtract 3 from both sides.

36. Toby is going to buy a car. The total purchase price of the car is represented by variable C. He will pay D dollars immediately, and then he will make equal payments (P) each month for a certain number of months. Which equation below represents the amount of his monthly payment (P)?
    A. $\frac{C-D}{M}$
    B. $\frac{C}{M} - D$
    C. $\frac{M}{C-D}$
    D. $D - \frac{C}{M}$
    E. $\frac{C}{M}$

37. A baseball team sells T-shirts and sweatpants to the public for a fundraising event. The total amount of money the team earned from these sales was $850. Variable *t* represents the number of T-shirts sold and variable *s* represents the number of sweatpants sold. The total sales in dollars is represented by the equation 25*t* + 30*s*. The amount earned by selling sweatpants is what fraction of the total amount earned?
    A. *s*/850
    B. 30*s*/850
    C. (25*t* + 30*s*)/850
    D. *t*/850
    E. 25*t*/850

38. The total funds, represented by variable F, available for P charity projects is represented by the equation F = $500P + $3,700. If the charity has $40,000 available for projects, what is the greatest number of projects that can be completed?
    A. 72
    B. 73
    C. 74

D. 79
E. 80

39. If $3x - 9 = -18$, then $x = ?$
   A. −6
   B. 6
   C. −3
   D. 3
   E. 2

40. Read the information below and answer the question that follows.

   - If door A is locked with the red key, then door B is locked with the blue key.
   - If door B is locked with the blue key, then door C is locked with the green key.
   - The key that locks door B also locks door D.

   If door A is locked with the red key, then which of the following must be true?
   A. Door C is locked with the blue key.
   B. Door C is locked with the red key.
   C. Door D is locked with the blue key.
   D. Door D is locked with the red key.
   E. Door B is currently unlocked.

41. The area of a square floor is 64 square units. The floor needs to be covered entirely with tiles. Each floor tile is 4 square units. How many tiles are needed to cover the floor?
   A. 8
   B. 12
   C. 16
   D. 24
   E. 36

42. Which of the following dimensions would be needed in order to find the area of the figure?

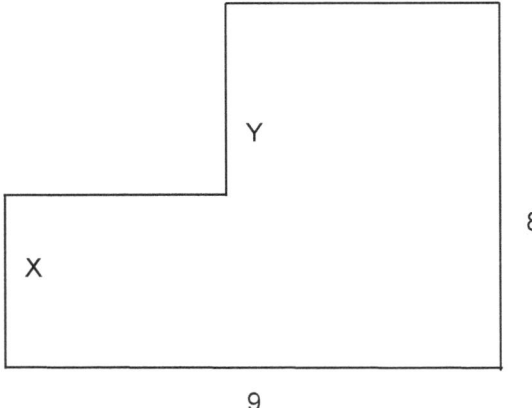

   A. X only
   B. Y only
   C. Both X and Y

D. Either X or Y
E. Neither X nor Y

43. Brooke wants to put new flooring in her living room. She will buy the flooring in square pieces that measure 1 square foot each. The entire room is 12 feet by 16 feet. The bookcases are one and a half feet deep from front to back. Flooring will not be put under the bookcases. Each piece of flooring costs $4.25. A diagram of her living room is provided.

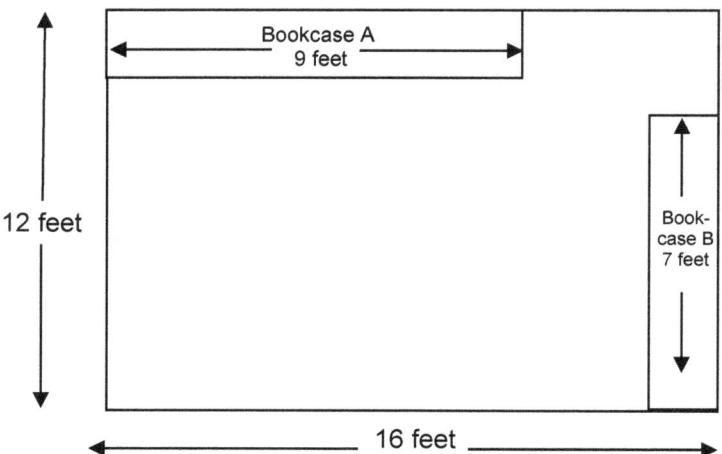

How much will Brooke pay to cover her living room floor? Provide the cost to the nearest dollar.
A. 168
B. 192
C. 714
D. 758
E. 816

44. A football field is 100 yards long and 30 yards wide. What is the area of the football field in square yards?
A. 3000
B. 1500
C. 300
D. 260
E. 240

45. A small pasture has a length of 5 yards and a width of 3 yards. Barbed wire will be placed on all four sides of the outside of this pasture. How many yards of barbed wire should be ordered?
A. 15
B. 16
C. 18
D. 40
E. 42

46. The graph below shows the relationship between the total number of hamburgers a restaurant sells and the total sales in dollars for the hamburgers. What is the sales price per hamburger?

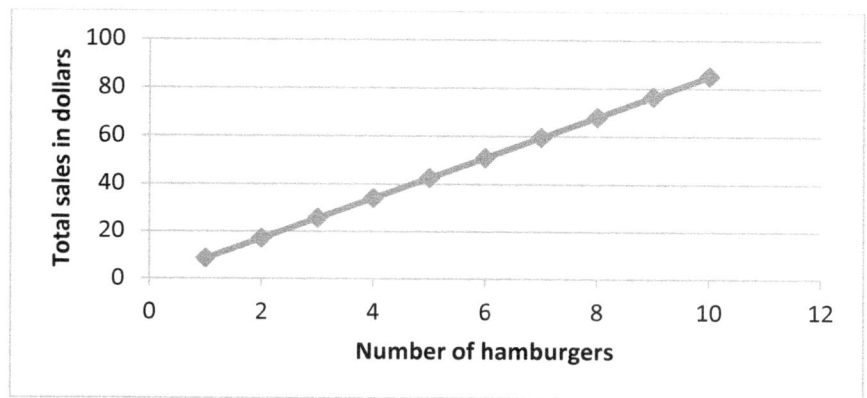

A. $4.00
B. $8.00
C. $8.50
D. $9.50
E. $10.00

47. In Brown County Elementary School, parents are advised to have their children vaccinated against five childhood diseases. According to the chart below, how many children were vaccinated against at least three diseases?

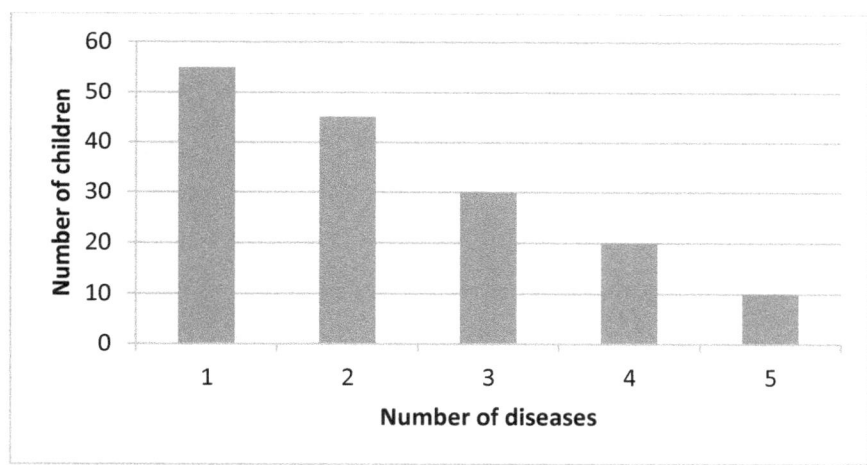

A. 30
B. 50
C. 60
D. 100
E. 130

48. A zoo has reptiles, birds, quadrupeds, and fish. At the start of the year, they have a total of 1,500 creatures living in the zoo. The pie chart below shows percentages by category for the 1,500 creatures at the start of the year.

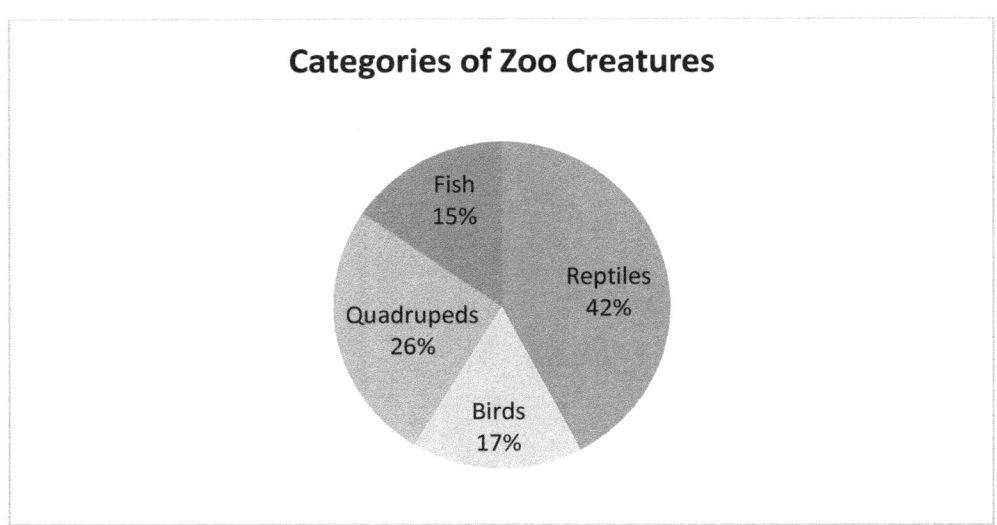

At the end of the year, the zoo still has 1,500 creatures, but reptiles constitute 40%, birds 23%, and quadrupeds 21%. How many more fish were there at the end of the year than at the beginning of the year?
A. 10
B. 11
C. 15
D. 16
E. 150

49. The students at Lyndon High School have been asked about their plans to attend the Homecoming Dance. The chart below shows the responses of each grade level by percentages. Which figure below best approximates the percentage of the total number of students from all four grades who will attend the dance? Note that each grade level has roughly the same number of students.

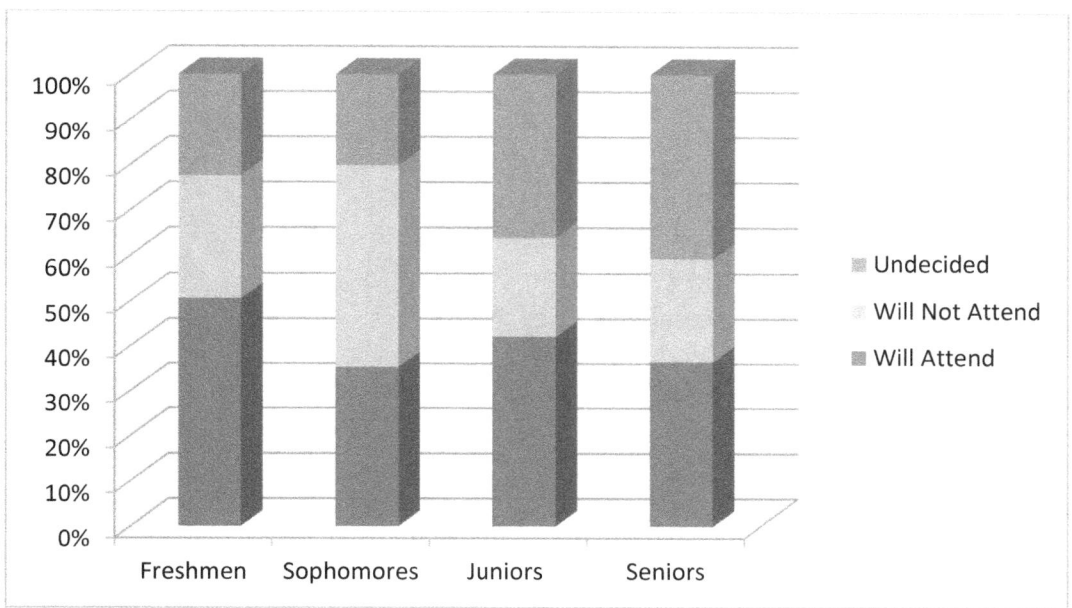

A. 25%
B. 35%
C. 45%
D. 55%
E. 60%

50. Look at the pie chart below and answer the question that follows.

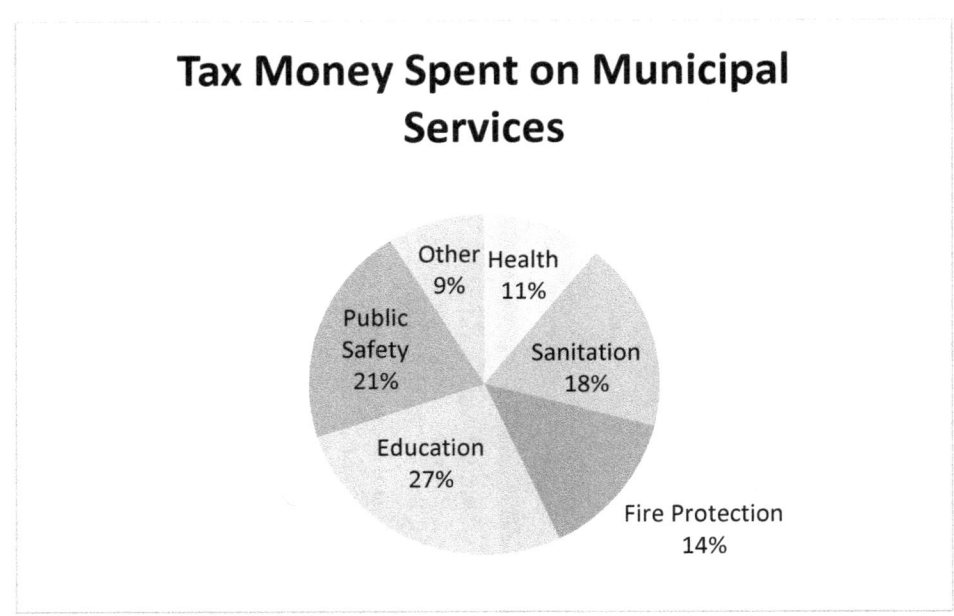

For next year, $6,537,200 in total tax money is budgeted for all municipal services. Each category is allocated the same percentage of next year's budget as the actual percentage spent for the current year. What is the budget amount for public safety?
A. $915,208
B. $1,107,813
C. $1,372,812
D. $1,765,004
E. $1,765,040

CBEST Math Practice Test 4 – Answers

1. The correct answer is A.
   In order to answer questions on ordering numbers from least to greatest or greatest to least, remember these principles: Negative numbers are less than positive numbers; (b) When two fractions have the same numerator, the fraction with the smaller number in the denominator is the larger According to these principles, $-1/4$ is less than $1/8$, $1/8$ is less than $1/6$, and $1/6$ is less than 1.

2. The correct answer is B.
   Perform the division, and then check the decimal places of the numbers. Divide as follows: 1523.48 ÷ 100 = 15.2348. Reading our result from left to right: 1 is in the tens place, 5 is in the ones place, 2 is in the tenths place, 3 is in the hundredths place, 4 is in the thousandths place, and 8 is in the ten-thousandths place.

3. The correct answer is E.
   STEP 1: Calculate the average high temperature in Fahrenheit. To do so, find the total for all five days and divide the result by 5: (72 + 68 + 65 + 82 + 81) ÷ 5 = 368 ÷ 5 = 73.6°F average.
   STEP 2: Convert the average in Fahrenheit to Celsius using the formula. To convert Fahrenheit to Celsius, we use this formula: °C = 0.56(°F − 32) = 0.56(73.6° − 32) = 0.56(41.6) = 0.56 × 41.6 = 23.296, which we round to 23°C.

4. The correct answer is B.
   Divide the total amount by the sales price per unit to solve: $7,375 ÷ $59 = 125 units sold

5. The correct answer is D.
   First, we can perform division to determine that the plane travels 6.5 miles per minute: 780 miles ÷ 120 minutes = 6.5 miles per minute. Since the plane travels at a constant speed, we multiply this amount by 40 minutes to solve: 6.5 miles per minute × 40 minutes = 260 miles

6. The correct answer is D.
   STEP 1: Determine the amount of time in seconds: 2 minutes and 48 seconds = (2 minutes × 60 seconds per minute) + 48 seconds = 120 seconds + 48 seconds = 168 seconds.
   STEP 2: Divide by the number of furlongs to find the rate: 168 seconds ÷ 12 furlongs = 14 seconds per furlong

7. The correct answer is D.
   The problem uses the phrase '2 out of every 20 employees' so we know that there are 2 employees who form a subset within each group of 20. STEP 1: Take the total number of employees and divide this by 20: 480 ÷ 20 = 24. STEP 2: Take the result from Step 1 and multiply by the amount in the subset to solve: 24 × 2 = 48

8. The correct answer is A.
   STEP 1: Add the charge for postage and handling to the original price per item: $22 + $3 = $25.
   STEP 2: Take the result from Step 1 and multiply by the number of items sold: $25 × 32 = $800

9. The correct answer is C.
   Divide and then express the result as a percentage. STEP 1: Treat the line in the fraction as the division symbol: 6/25 = 6 ÷ 25 = 0.24. STEP 2: To express the result from Step 1 as a percentage, move the decimal point two places to the right and add the percent sign: 0.24 = 24.0%

10. The correct answer is C.
    Add the four figures together to solve: 163.75 + 107.50 + 91.25 + 10.30 = 372.80

11. The correct answer is B.
    STEP 1: Find the amount of material needed for each quilt: 2 yards red, 4 yards blue, 1.2 yards gold (6 ÷ 5 = 1.2), and 0.5 yards white = 2 + 4 + 1.2 + 0.5 = 7.7 yards each. STEP 2: Multiply the total number of quilts by the number of yards per quilt to solve: 10 × 7.7 = 77
12. The correct answer is D.
    Convert the cups to quarter cups: 10 cups ÷ ¼ = 40 quarter cups. Then combine the whole number with the fraction and multiply to solve: $40^{1}/_{4}$ × 50 cents per quarter cup = 40.25 × 0.50 = $20.50
13. The correct answer is D.
    The problem is asking for the total for all five months, so round and then add the amounts together to solve: $700 + $600 + $600 + $800 + $1,000 = $3,700
14. The correct answer is D.
    STEP 1: Add the whole numbers: 4 + 3 = 7. STEP 2: Add the fractions: 3/8 + 7/8 = 10/8. STEP 3: Simplify the fraction from Step 2: 10/8 = 8/8 + 2/8 = 1 + 2/8 = $1^{2}/_{8}$ = $1^{1}/_{4}$. STEP 4: Combine the results from Step 1 and Step 3 to solve the problem: 7 + $1^{1}/_{4}$ = $8^{1}/_{4}$
15. The correct answer is C.
    In this problem, the fraction on the second number is larger than the fraction on the first number, so we need to convert the first fraction before we start our calculation. STEP 1: Convert $12^{7}/_{16}$ for subtraction: $12^{7}/_{16}$ = $11^{7}/_{16}$ + 1 = $11^{7}/_{16}$ + $^{16}/_{16}$ = $11^{23}/_{16}$. STEP 2: There were $8^{9}/_{16}$ yards left, so subtract the whole numbers: 11 − 8 = 3. STEP 3: Subtract the fractions: 23/16 − 9/16 = 14/16. STEP 4: Simplify the fraction from Step 3: 14/16 = (14 ÷ 2)/(16 ÷ 2) = 7/8. STEP 5: Combine the results from Step 2 and Step 4 to get your new mixed number to solve the problem: 3 + 7/8 = $3^{7}/_{8}$
16. The correct answer is B.
    The problem tells us the relative amount of units sold, but the question is asking for the relative amount of units left. So, subtract the decimal from 1 to find the relative amount left: 1 − 0.75 = 0.25. Then multiply the total number of items at the start by this decimal number: 80 items × 0.25 = 80 × 0.25 = 20 items left
17. The correct answer is B.
    STEP 1: Work out the cost for the usual supplier: 325 pairs × $4 = $1,300. STEP 2: Calculate the price for the second supplier: $1,250 + ($1,250 × .06) = $1,250 + $75 = $1,325. STEP 3: Compare to the third deal to solve: The third deal is $1,290 so this is the best deal.
18. The correct answer is C.
    STEP 1: Take the total number of employees and divide this by 5: 250 ÷ 5 = 50. STEP 2: The question asks how many questionnaires have not been completed and returned, so subtract to find the amount in the 'not returned' subset: 5 − 4 = 1. STEP 3: Multiply the result from step 2 by the result from step 1 to solve: 50 × 1 = 50
19. The correct answer is D.
     STEP 1: Determine the total for sales in December: $20 × 55 = $1,100. STEP 2: Determine the total sales for January: $12 × 20 = $240. STEP 3: Add these two amounts to solve: $1,100 + $240 = $1,340
20. The correct answer is E.
    He needs to subtract the 92 that he added by mistake to get back to his starting point. Then he needs to subtract 92 again to get the correct result. So, he can subtract 92 two times or simply shortcut by subtracting 184 (92 + 92 = 184).

21. The correct answer is D.
    The question is asking us about a time duration of 6 minutes, so we need to calculate the amount of seconds in 6 minutes: 6 minutes × 60 seconds per minute = 360 seconds in total. Then divide the total time by the amount of time needed to make one journey: 360 seconds ÷ 45 seconds per journey = 8 journeys. Finally, multiply the number of journeys by the amount of inches per journey in order to get the total inches: 10.5 inches for 1 journey × 8 journeys = 84 inches in total
22. The correct answer is B.
    For your first step, add the subsets of the ratio together: 6 + 7 = 13. Then divide this into the total: 117 ÷ 13 = 9. Finally, multiply the result from the previous step by the subset of males from the ratio: 6 × 9 = 54 males in the class
23. The correct answer is D.
    Be careful with the phrase "out of" in proportion questions like this one. We are given the phrase "4 out of 7." If 4 out of 7 prefer the toffee flavor, then the remaining 3 prefer the coffee flavor. Accordingly, the ratio of flavor preference of toffee to coffee is 4 to 3. So, after determining the ratio, your next step is to divide the total by 7: 1540 7 = 220. Then multiply this by 3 for the coffee-flavor preference: 220 3 = 660.
24. The correct answer is C.
    STEP 1: Determine the commission earned per hour: $15 charged − $12 paid to employee = $3 per hour commission. STEP 2: Calculate the total hours that the 10 employees worked: 10 × 40 = 400 hours in total. STEP 3: Multiply the total number of hours by the commission per hour to solve: 400 hours × $3 commission per hour = $1,200 total commission
25. The correct answer is D.
    Divide to solve: 49 ÷ 50 = 0.98 = 98%
26. The correct answer is D.
    Calculate the total, and divide by the number of days. STEP 1: Find the total: $90 + $85 + $85 + $105 + $110 = $475. STEP 2: Divide the result from Step 1 by the number of days: $475 ÷ 5 = $95
27. The correct answer is D.
    STEP 1: Add the whole numbers: 8 + 7 = 15. STEP 2: Add the fractions: 3/4 + 1/2 = 3/4 + 2/4 = 5/4. STEP 3: Simplify the fraction from Step 2: 5/4 = 4/4 + 1/4 = $1^{1}/_{4}$ = 1 foot and 3 inches. STEP 4: Combine the results from Step 1 and Step 3 to solve the problem: 15 feet + 1 foot and 3 inches = 16 feet and 3 inches
28. The correct answer is C.
    In this problem, the fraction on the second number is larger than the fraction on the first number, so we need to convert the first fraction before we start our calculation. STEP 1: Convert $28^{3}/_{10}$ for subtraction: $28^{3}/_{10}$ = $27^{3}/_{10}$ + 1 = $27^{3}/_{10}$ + $^{10}/_{10}$ = $27^{13}/_{10}$. STEP 2: Subtract the whole numbers. $7^{9}/_{10}$ hours have been spent on the job so far, so subtract the 7 hours: 27 − 7 = 20. STEP 3: Subtract the fractions: 13/10 − 9/10 = 4/10. STEP 4: Simplify the fraction from Step 3: 4/10 = (4 ÷ 2)/(10 ÷ 2) = 2/5. STEP 5: Combine the results from Step 2 and Step 4 to get your new mixed number to solve the problem: 20 + 2/5 = $20^{2}/_{5}$
29. The correct answer is C.
    First of all, convert 96 yards to feet: 36 yards × 3 feet in a yard = 108 feet. Then subtract this from 116 feet: 116 − 108 = 8 feet. Then convert the feet to inches for your answer: 8 feet × 12 inches per foot = 96 inches

30. The correct answer is D.
    In order to calculate the average, you simply add up the values of all of the items in the set, and then divide by the number of items in the set. To solve problems like this one, set up an equation to calculate the average, using x for the unknown value. Set up your equation to calculate the average, using x for the age of the 5th sibling:
    $(2 + 5 + 7 + 12 + x) \div 5 = 8$
    $(2 + 5 + 7 + 12 + x) \div 5 \times 5 = 8 \times 5$
    $(2 + 5 + 7 + 12 + x) = 40$
    $26 + x = 40$
    $26 - 26 + x = 40 - 26$
    $x = 14$

31. The correct answer is D.
    3 tablespoons of cocoa powder and ¼ teaspoon of baking powder are needed for the original recipe to make 4 brownies. There are 3 teaspoons in a tablespoon, so calculate the total teaspoons needed for the original recipe first: 3 tablespoons × 3 = 9 teaspoons cocoa powder + ¼ teaspoon baking powder = 9¼ teaspoons in total. We are now making 12 brownies, so we need to multiply all of the ingredients by 3: 9¼ × 3 = 27¾ teaspoons

32. The correct answer is B.
    First add up all of the numbers to get the total: 35 + 42 + 23 + 20 = 120. We divide the number of office workers who live in the country, which is 20 people, into this total to get our answer. We can find this amount in the lower right-hand corner of the chart. So, set up the fraction and then simplify: $20/120 = 2/12 = 1/6$

33. The correct answer is B.
    First, we will use variable T as the total number of items in the set. The probability of getting a red scarf is $1/3$. So, set up an equation to find the total items in the data set.
    $$\frac{5}{T} = \frac{1}{3}$$
    $$\frac{5}{T} \times 3 = \frac{1}{3} \times 3$$
    $$\frac{5}{T} \times 3 = 1$$
    $$\frac{15}{T} = 1$$
    $$\frac{15}{T} \times T = 1 \times T$$
    $$15 = T$$
    We have 5 red scarves, 6 blue scarves, and x green scarves in the data set that make up the total sample space, so now subtract the amount of red and blue scarves from the total in order to determine the number of green scarves.
    $5 + 6 + x = 15$
    $11 + x = 15$
    $11 - 11 + x = 15 - 11$
    $x = 4$

34. The correct answer is C.
    Isolate the integers to one side of the equation.

$$\frac{3}{4}x - 2 = 4$$
$$\frac{3}{4}x - 2 + 2 = 4 + 2$$
$$\frac{3}{4}x = 6$$

Then get rid of the fraction by multiplying both sides by the denominator.

$$\frac{3}{4}x \times 4 = 6 \times 4$$

$3x = 24$

Then divide to solve the problem.

$3x \div 3 = 24 \div 3$

$x = 8$

35. The correct answer is A.
Remember that to solve problems like this, you need to deal with the integers (whole numbers) and then isolate the variable (x). The solution is as follows:
$4x - 3 = 2$
$4x - 3 + 3 = 2 + 3$ (Add 3 to both sides of the equation.)
$4x = 5$
$4x \div 4 = 5 \div 4$ (Divide both sides by 4.)
$x = \frac{5}{4}$

36. The correct answer is A.
The total amount that Toby has to pay is represented by C. He is paying D dollars immediately, so we can determine the remaining amount that he owes by deducting his down payment from the total. So, the remaining amount owing is represented by the equation: C – D. We have to divide the remaining amount owing by the number of months to get the monthly payment (P):
P = (C – D) ÷ M = $\frac{C-D}{M}$

37. The correct answer is B.
We need to set up a fraction, the numerator of which consists of the amount of sales in dollars for sweatpants, and the denominator of which consists of the total amount of sales in dollars for both items. The problem tells us that the amount of sales in dollars for sweatpants is 30*s* and the total amount of sales is 850, so the answer is 30*s*/850.

38. The correct answer is A.
The equation is: F = $500P + $3,700. We are told that the total funds are $40,000 so put that in the equation to solve the problem.
$40,000 = $500P + $3,700
$40,000 – $3,700 = $500P
$36,300 = $500P
$36,300 ÷ 500 = $500 ÷ 500P
$36,300 ÷ 500 = 72.6

39. The correct answer is C.
Add 9 to each side to get rid of the integer on the left.
$3x - 9 = -18$
$3x - 9 + 9 = -18 + 9$
$3x = -9$
Then divide each side by 3 to solve.

$3x ÷ 3 = -9 ÷ 3$
$x = -3$

40. The correct answer is C.
    If door A is locked with the red key, then door B is locked with the blue key. If the key that locks door B also locks door D, then the blue key is also used to lock door D.

41. The correct answer is C.
    Since we are dealing with a square, all four sides of the floor are equal to each other. The tiles are also square, so they also have equal sides. Therefore, we can simply divide to get the answer: $64 ÷ 4 = 16$

42. The correct answer is C.
    To solve problems like this one, try to visualize two rectangles. The first rectangle would measure 8 × 9 and the second rectangle would measure X × Y. Essentially a rectangle is missing at the upper left-hand corner of the figure. We would need to know both the length and width of the "missing" rectangle in order to calculate the area of our figure. So, we need to know both X and Y in order to solve the problem.

43. The correct answer is C.
    Calculate the total square footage of the room first: $12 × 16 = 192$ total square feet. Then calculate the area under each bookcase. Bookcase A: $9 × 1.5 = 13.5$; Bookcase B: $7 × 1.5 = 10.5$; Total bookcase area: $13.5 + 10.5 = 24$ square feet. Next find the area of the room without the bookcase area: $192 - 24 = 168$ square feet to be covered. Finally, determine the cost: 168 square feet × $4.25 per piece = $714 total cost

44. The correct answer is A.
    The area of a rectangle is equal to its length times its width. This football field is 30 yards wide and 100 yards long, so we can substitute the values into the appropriate formula.
    rectangle area = width × length
    rectangle area = 30 × 100
    rectangle area = 3000

45. The correct answer is B.
    You are being asked about the distance around the outside, so you need to calculate the perimeter, Write out the formula: (length × 2) + (width × 2). Then substitute the values: (5 × 2) + (3 × 2) = 10 + 6 = 16

46. The correct answer is C.
    For ten hamburgers, the total price is $85, so each hamburger sells for $8.50: $85 total sales in dollars ÷ 10 hamburgers sold = $8.50 each

47. The correct answer is C.
    The quantity of diseases is indicated on the bottom of the graph, while the number of children is indicated on the left side of the graph. To determine the number of children that have been vaccinated against three or more diseases, we need to add the amounts represented by the bars for 3, 4, and 5 diseases: $30 + 20 + 10 = 60$ children

48. The correct answer is C.
    At the beginning of the year, 15% of the 1,500 creatures were fish, so there were 225 fish at the beginning of the year ($1{,}500 × 0.15 = 225$). In order to find the percentage of fish at the end of the year, we first need to add up the percentages for the other animals: $40\% + 23\% + 21\% = 84\%$. Then subtract this amount from 100% to get the remaining percentage for the fish: $100\% - 84\% = 16\%$. Multiply the percentage by the total to get the number of fish at the end of the year:

1,500 × 0.16 = 240. Then subtract the beginning of the year from the end of the year to calculate the increase in the number of fish: 240 − 225 = 15

49. The correct answer is B.
The dark gray part at the bottom of each bar represents those students who will attend the dance. 45% of the freshman, 30% of the sophomores, 38% of the juniors, and 30% of the seniors will attend. Calculating the average, we get the overall percentage for all four grades:
(45 + 30 + 38 + 30) ÷ 4 = 35.75%
35% is the closest answer to 35.75%, so it best approximates our result.

50. The correct answer is C.
Take the total dollar amount of the budget and multiply by the 21% for public safety: $6,537,200 × 0.21 = $1,372,812

CBEST READING

Format of the CBEST Reading Test

Two types of reading skills are assessed on the CBEST reading test:

(1) Critical Analysis and Evaluation

(2) Comprehension and Research

<u>Type 1 Questions</u>

Critical analysis and evaluation questions assess your understanding of the specific details within a reading selection or selections.

Critical analysis and evaluation questions also cover understanding the author's purpose, technique, or assumptions, as well as making comparisons and predictions.

<u>Type 2 Questions</u>

Comprehension questions cover the logical organization of the reading selection.

This means that you will see questions on the relationship of ideas within a text, as well as having to identify the main idea or draw inferences.

Research questions assess understanding tables of contents, as well as comprehending data that is represented in tables, charts, or graphs.

Types of Questions on the CBEST Reading Test

You will see these specific types of questions on the CBEST reading test:

(1) CRITICAL ANALYSIS AND EVALUATION
- Comparing and contrasting ideas within a reading selection or selections
- Identifying details that support the author's main idea
- Predicting an outcome based on information from a selection
- Understanding the author's viewpoint or attitude
- Determining the relevance of ideas to a selection
- Recognizing arguments for or against a viewpoint
- Understanding the author's persuasive techniques or strategies
- Identifying the author's purpose or assumptions
- Discerning facts from opinions in a selection
- Identifying inconsistencies or differences between two or more paragraphs or selections
- Recognizing the intended audience for a selection
- Identifying the tone or style of writing

(2) COMPREHENSION AND RESEARCH SKILLS
- Identifying and understanding relationships between general and specific ideas in a selection
- Determining the correct order of events or of steps in a process
- Arranging ideas logically within a selection
- Recognizing the main idea of a selection

- Paraphrasing and summarizing ideas in selections
- Drawing conclusions and making generalizations
- Recognizing implications and making inferences
- Determining the meaning of unknown words or metaphorical phrases
- Identifying varying interpretations of words or information in a selection
- Understanding the relationship between meaning and context
- Using key transition and linking words correctly within a selection
- Understanding the organizational schemes of various selections
- Understanding how to use the table of contents or index of a selected book to locate information
- Identifying the place where specific information is located within a book
- Using and interpreting information in tables, graph, or charts

How to Use the Reading Study Guide

The reading practice tests in this study guide contain questions of all of the types that you will see on the real CBEST test.

Practice test 1 in this book is in "tutorial mode."

As you complete practice test 1, you should pay special attention to the tips highlighted in the special boxes.

Although you will not see tips like this on the actual exam, these suggestions will help you improve your performance on each subsequent practice test in this publication.

You should also study the explanations to the answers to practice test 1 especially carefully.

Studying the tips and explanations in reading practice test 1 will help you obtain strategies to improve your performance on the other practice tests in this book.

Of course, these strategies will also help you do your best on the day of your actual CBEST test.

CBEST Practice Reading Test 1

Look at the extract from an index below to answer the two questions that follow.

```
Light  20–45
       beams  29
       bulbs  30–41
            energy-saving  40–41
            florescent  32–33
            halogen  36–39
            incandescent  30–31
            neon  34–35
       emission of  20–23
       emitting diode  27, 29
       speed of  24–26
            measuring  24–25
            physics  26
       waves  28
Lightning  93–96
       causes of  93
       surviving strikes of  95
```

1. What method best describes the organization of the section on light bulbs?
   A. by date
   B. by popularity
   C. by energy consumption
   D. by cost per bulb
   E. by type

2. Where can the reader look to see whether the book contains information on how the emission of light is measured?
   A. Pages 27 and 29
   B. Pages 20–23
   C. Pages 24–25
   D. Page 26
   E. Page 28

> Questions 1 and 2 are research skills questions.
> Question 1 assesses your understanding of how to use an index, while question 2 covers the skill of identifying where specific information is located within a book.

Tips and Explanations:

1. The correct answer is E. Using an index or table of contents correctly requires you to understand the organizational scheme of the sections within the table of contents or index. You need to look at the information in each section and then ask yourself what the items within the section have in common. We can see that the section on light bulbs is organized by type of light bulb because energy-saving, florescent, halogen, incandescent, and neon are different kinds of light bulbs.

2. The correct answer is B. When questions ask you to identify where specific information is located within a book, you need to read through each part of the table of contents or index. The indentation of each section is extremely important for these types of questions. For instance, look at this excerpt from the index:

> Lightning 93–96
>   causes of 93

The indented item always relates to the category above. So, the causes of lighting are discussed on page 93. Our question is asking us about the measurement of the emission of light. Based on the indentations, the reader can look at pages 20 to 23 to see whether the book contains information on how the emission of light is measured. Remember to be very careful when considering the indentation in the index. Pages 20–23, on line 9 of the index, deal with the emission of light, although if you misread the index, line 9 looks like it deals with the emission from bulbs.

Read the passage below and answer the four questions that follow.

Research shows that the rise in teenage smoking over the last ten years took place primarily in youth from more affluent families, in other words, families in which both parents were working and earning good incomes. Therefore, these teenagers were not from disadvantaged homes, as most people seemed to believe.

The facts demonstrate quite the opposite because the most striking and precipitous rise in smoking has been for teenagers from the most financially advantageous backgrounds. Furthermore, because of various lawsuits against the major tobacco companies, the price of cigarettes has actually declined sharply over the past decade. The paradox is that the increased demand for cigarettes originated from new teenage smokers who were from well-off families. Yet, contrary to these market forces, the price of tobacco products fell during this time.

3. What is the primary purpose of this passage?
    A. to provide information on a recent trend
    B. to emphasize the dangers of smoking
    C. to dispel a common misconception
    D. to highlight the difference between two types of teenagers
    E. to criticize teenage smokers

4. Which of the following is the best meaning of the word precipitous as it is used in this passage?
    A. unreliable
    B. unbelievable
    C. predictable
    D. dangerous
    E. dramatic

5. From this passage, it seems safe to conclude which of the following?
    A. The majority of new teenage smokers in the last ten years could have afforded to pay higher prices for tobacco.
    B. Parents of affluent families are often not aware of the smoking habits of their children.
    C. Smoking among teenagers from disadvantaged homes also increased during the past decade.
    D. Major tobacco companies have recently faced bankruptcy.
    E. There has been a rise in smoking cessation programs for teenagers during the past decade.

6. Which of the following statements gives the best summary of the main points of the lecture?
   A. Teenagers from affluent families smoke more than teenagers from disadvantaged homes.
   B. The price of tobacco products is normally unrelated to market forces.
   C. The price of cigarettes has fallen more than expected during the last ten years.
   D. Contrary to popular belief, the rise in teenage smoking during the last ten years has been attributable to youth from wealthy family backgrounds.
   E. There has been a noticeable increase in teenage smoking in recent years.

> Question 3 is a critical analysis question on the author's purpose.
> Question 4 is a comprehension question on the meaning of unknown words.
> Question 5 is a comprehension question on drawing conclusions.
> Question 6 is an example of a question that asks you to summarize information from a passage.

Tips and Explanations:

3. The correct answer is C. In order to determine the author's purpose, you need to pay special attention to the last sentence of the first paragraph of a selection. This is where the writer normally puts his or her thesis statement, which is the author's main assertion or primary purpose. Looking at the last sentence of the first paragraph, we can see that the primary purpose of this passage is to dispel a common misconception. The idea of a misconception is indicated in the phrase "as most people seem to believe" from the last sentence of the first paragraph.

4. The correct answer is E. For questions on the meaning of words, you need to look for other words in the passage that are synonyms for the word in the question. "Dramatic" and "precipitous" are synonyms in the context of this passage. The word "striking" from the first sentence of the second paragraph of the passage is also a synonym for "precipitous."

5. The correct answer is A. When drawing conclusions, look for words and phrases in the passage that express the writer's viewpoint. See the phrase "contrary to these market forces" in the last sentence of the passage. The market forces refer to the factors that would have caused the price of tobacco to increase. Based on the word "contrary," it seems safe to conclude that the majority of new teenage smokers in the last ten years could have afforded to pay higher prices for tobacco, but in spite of this fact, the price did not go up.

6. The correct answer is D. For questions asking you to summarize the main points, you first must identify what the main points are. Paragraph 1 states that there has been a recent misconception about teenage smoking. Paragraph 2 explains that the price of tobacco should have gone up because youngsters from wealthy families could have afforded to pay a higher price. The following statement gives the best summary of the main points of the lecture because it mentions the misconception, as well as the pricing aspect: "Contrary to popular belief, the rise in teenage smoking in the last ten years has been attributable to youth from wealthy family backgrounds." So, answer D is the best summary of this paragraph.

Read the passage below and answer the three questions that follow.

Over the past five years, sales of organic products in the United States have increased a staggering 20 percent, with retail sales per year of more than 9 billion dollars. American farmers have realized that organic farming is an incredibly cost-effective method because it can be used to control costs, _____ to appeal to higher-priced markets.

Organic farming has become one of the fastest growing trends in agriculture recently not only for monetary, but also for environmental reasons. _____ the monetary benefits, organic farming also results in positive ecological outcomes. That is because the use of chemicals and synthetic materials is strictly prohibited.

7. Which of the words or phrases, if inserted in order into the blanks in the passage, would help the reader better understand the sequence of events?
   A. besides; As a result of
   B. as well as; Apart from
   C. while; In addition to
   D. in addition; However
   E. but also; Considering

8. Which sentence or phrase from the passage best expresses its central idea?
   A. Over the past five years, sales of organic products in the United States have increased a staggering 20 percent, with retail sales per year of more than 9 billion dollars.
   B. American farmers have realized that organic farming is an incredibly cost-effective method because it can be used to control costs.
   C. Organic farming has become one of the fastest growing trends in agriculture recently not only for monetary, but also for environmental reasons.
   D. Organic farming also results in positive ecological outcomes.
   E. That is because the use of chemicals and synthetic materials is strictly prohibited.

9. What word best describes the style of writing in this passage?
   A. commercial
   B. technical
   C. scientific
   D. explanatory
   E. polemical

> Question 7 is a comprehension question on using key transition and linking words.
> Question 8 is a comprehension question on recognizing the main idea.
> Question 9 is an example of a question on identifying the style of writing in a selection.

Tips and Explanations:

7. The correct answer is B. For questions on transition and linking words, look carefully at the information that is provided in the selection after the gap. Determine whether an additional supporting idea is being stated or if the author is changing the subject. Then look at the answer choices to see which one matches the flow of the text. "As well as" is the most suitable answer for the first gap in this selection because another reason is being given. "Apart from" fits best into the second gap because ecological outcomes are mentioned, in addition to the environmental reasons.

8. The correct answer is C. For main idea questions, look to see which ideas are stated in each part of the selection. In the selection above, the first half of the passage addresses economics, while the second half talks about the environment. The central idea of the passage is therefore that organic farming has become one of the fastest growing trends in agriculture recently not only for monetary, but also for environmental reasons. This is the only answer that expresses both of the ideas. The other answers are merely restating specific points from the selection.

9. The correct answer is D. In order to identify the style of a selection, you need to examine the transitional words and phrases in the passage. We can see that reasons and explanations are given in the passage, using the phrases "not only," "but also," and "that is because." Accordingly, the writing style is explanatory.

Read the passage below and answer the three questions that follow.

The corpus of research on Antarctica has resulted in an abundance of factual data. For example, we now know that more than ninety-nine percent of the land is completely covered by snow and ice, making Antarctica the coldest continent on the planet. This inhospitable climate has brought about the adaptation of a plethora of plants and biological organisms present on the continent. Investigations into the sedimentary geological formations provide testimony to the process of adaptation. Sediments recovered from the bottom of Antarctic lakes, as well as bacteria discovered in ice, have been of invaluable significance because they have revealed the history of global climate change over the past 10,000 years.

10. According to the passage, the plants and organisms in Antarctica:
    A. have survived because of the process of adaptation.
    B. are the result of sedimentary geological formations.
    C. cover more than 99% of the land surface.
    D. grow in the bottom of lakes on the continent.
    E. reveal the history of climate change over the past 10,000 years.

11. The information the writer conveys in this passage is addressed mainly to:
    A. professional geologists.
    B. tourists taking part in a trip to Antarctica.
    C. elementary school children.
    D. researchers prior to an expedition.
    E. students attending a college lecture.

12. Which one of the following statements best expresses the writer's opinion regarding the corpus of research on Antarctica?
    A. It was exceptionally difficult due to the snow and ice coverage on the ground.
    B. It provides us with more accurate information on colder climates.
    C. It has helped us better to understand historical climatic fluctuations in the rest of the world.
    D. It reveals important data on organisms typically found in lakes.
    E. It is linked to changes in world history.

> For the selection above, we have three types of critical analysis and evaluation questions.
> Question 10 is a critical analysis question on identifying details from a passage.
> Question 11 is a critical analysis question that asks you to recognize the intended audience for the selection.
> Question 12 is an evaluation asking you to understand the author's viewpoint.

Tips and Explanations:

10. The correct answer is A. For questions on identifying details from the passage, you need to locate precise information within the selection. You can do this by using keywords. Looking at question 10, we can see that we are required to find specific details about plants and organisms. The keywords

"plants" and "organisms" are mentioned in the third sentence of the passage, which states: "This inhospitable climate has brought about the adaptation of a plethora of plants and biological organisms present on the continent." Answer A also mentions the process of adaption, so it is the correct answer.

11. The correct answer is E. When you have to identify the intended audience, you should evaluate the vocabulary that the passage uses. In this selection, we see words such as "research," "data," and "testimony." Then ask yourself: Who uses this type of vocabulary? Words like these are too academic for tourists or children, but not technical enough for professionals or researchers. Therefore, we know that the audience for selection consists of students attending a college lecture.

12. The correct answer is C. In order to understand the author's viewpoint, you need to look for adjectives. Words such as "important," "helpful," or "significant" will be in the selection if the writer has a positive viewpoint. If the writer is expressing a negative viewpoint, you will see adjectives such as "pointless" or "useless." In our selection, the answer is found in the last sentence of the passage in the phrase ". . . have been of invaluable significance because they have revealed the history of global climate change." The phrase "invaluable significance" reveals the author's positive view of the research, while the phrase "the history of global climate change" is synonymous with "historical climatic fluctuations in the rest of the world."

Read the passage below and answer the two questions that follow.

[1]The student readiness educational model is based on the view that students operate at different levels of ability. [2]For some students, this might mean that they are operating above the average ability level of their contemporaries. [3]Other students may be functioning at a level that is below average. [4]Of course, students in a particular class may be of slightly different ages. [5]There are also students who are at the optimum learning level. [6]The level is optimum because they are being challenged and learning new things, but they do not feel overwhelmed or inundated by the new information. [7]A teacher will engage students best with classroom learning activities that are fun and interesting.

13. Which numbered sentence provides an opinion rather than a fact?
    A. Sentence 1
    B. Sentence 2
    C. Sentence 5
    D. Sentence 6
    E. Sentence 7

14. Which numbered sentence is least relevant to the main idea of the passage?
    A. Sentence 2
    B. Sentence 3
    C. Sentence 4
    D. Sentence 5
    E. Sentence 6

> Question 13 is an evaluation question asking you to discern facts from opinions.
> Question 14 is another evaluation question. It requires you to determine the relevance of ideas to the selection.

Tips and Explanations:

13. The correct answer is E. Opinion questions are similar to viewpoint questions, which we have seen previously in question 12. For these types of questions, you need to look for adjectives that the

author uses. Here, we can see that sentence 7 is an opinion, not a fact. Sentence 7 states: "A teacher will engage students best with classroom learning activities that are fun and interesting." The adjective "best" indicates that an opinion is being given.

14. The correct answer is C. For relevancy questions, you should again look at the vocabulary. Try to determine if there is any word or idea that is being repeated in the majority of the sentences in the selection. In the selection above, we can see that the word "level" is used in the majority of sentences in the passage. Age is a different concept than level. So, sentence 4, which states that students in a particular class may be of slightly different ages, is least relevant to the main idea of the passage. In other words, we know from reading the passage that slight differences in ages should not affect the teaching or learning processes.

Use the graph below to answer the question that follows.

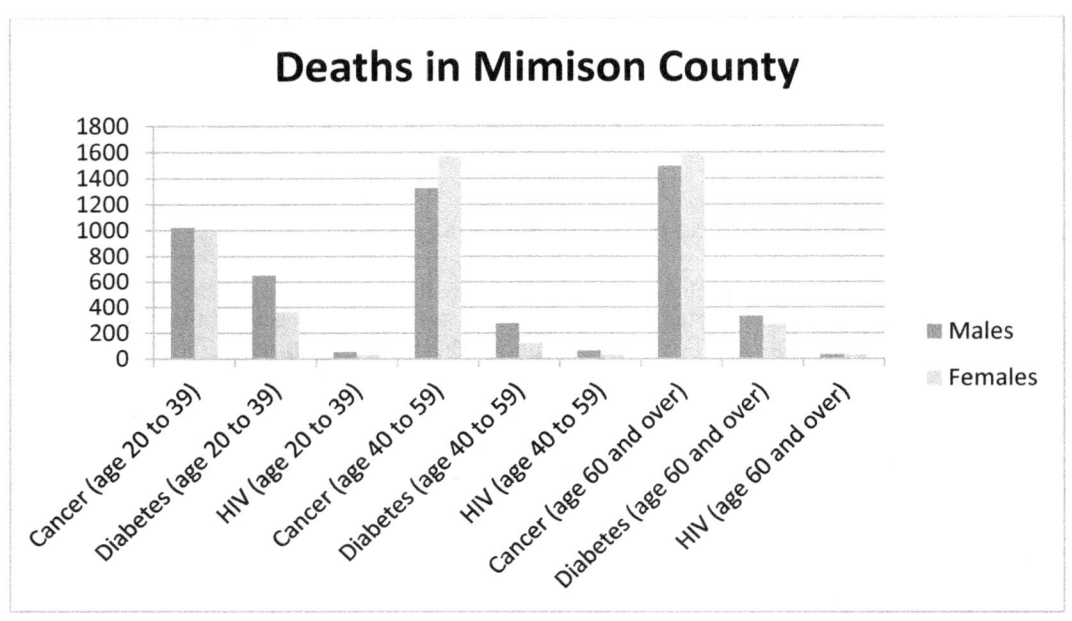

15. Which group in total had the highest number of deaths from all three diseases (cancer, diabetes, and HIV)?
    A. Males, age 20 to 39
    B. Females, age 20 to 39
    C. Males, age 40 to 59
    D. Females, age 40 to 59
    E. Females, age 60 and over

> Question 15 is a research skills question that requires you to interpret information from a chart or graph. In this case, we have to evaluate a bar graph.

Tips and Explanations:

15. The correct answer is E. For these types of questions, you need to look at the data in the chart or graph, and then visually evaluate the information for each group in total. For the graph above, first look at the groups of dark grey bars, which represent males. Then look at the light grey bars, which are for the female population. You will see that all of the light grey bars for the 40 to 59 group are slightly lower than those of the 60 and over group. We can also clearly see that the first light grey bar in the age 60 and over group is the highest one of any of the bars. Adding this first light grey bar for cancer at 1,600 to the 220 for diabetes and the 20 for HIV, we arrive at 1,840 in total for all three diseases for females over age 60, which is the highest number of deaths from all three diseases.

Read the passage below and answer the four questions that follow.

Earthquakes occur when there is motion in the tectonic plates on the surface of the earth. The crust of the earth contains twelve such tectonic plates. Fault lines, the places where these plates meet, build up a great deal of pressure because the plates are always pressing on each other. The two plates will eventually shift or separate because the pressure on them is constantly increasing, and this build-up of energy needs to be released. When the plates shift or separate, we have an occurrence of an earthquake, also known as a seismic event.

The point where the earthquake is at its strongest is called the epicenter. Waves of motion travel out from this epicenter, often causing widespread destruction to an area. With this likelihood for earthquakes to occur, it is essential that earthquake prediction systems are in place. _____
_____ . However, these prediction systems need to be more reliable in order to be of any practical use.

16. What happens immediately after the pressure on the tectonic plates has become too great?
    A. Fault lines are created.
    B. There is a build-up of energy.
    C. There is a seismic event.
    D. Waves of motion travel out from the epicenter.
    E. Prediction systems become more reliable.

17. Which sentence, if inserted into the blank line in paragraph 2, would best fit into the logical development of the passage?
    A. Unfortunately, many countries around the world do not have earthquake prediction systems.
    B. The purpose of earthquake prediction systems is to give advanced warning to the population, thereby saving lives in the process.
    C. Earthquakes can also occur at sea since there are tectonic plates in certain oceans.
    D. It is extremely costly to rectify the destruction that earthquakes cause.
    E. Aftershocks can also occur after the tremors of the earthquake have passed.

18. What inference about earthquakes can be drawn from the passage?
    A. There has been no discernible change in the number of earthquakes in recent years.
    B. There has been an increase in the destruction caused by earthquakes in recent years.
    C. The destruction of property could be avoided with improved earthquake prediction systems.
    D. The number of deaths from earthquakes could be lowered if earthquake prediction systems were more reliable.
    E. Earthquake prediction systems could help to lessen the strength of earthquakes.

19. The writer makes her final statement more compelling by preceding it by which of the following?
    A. a dispassionate, scientific explanation
    B. emotionally-evocative examples
    C. a historical account of events
    D. a prediction of a future catastrophe
    E. step-by-step instructions

> Question 16 is the type of comprehension question that asks you to determine the correct order of events or steps in a process.
> Question 17 is another type of comprehension question. It requires you to arrange ideas logically within a selection.
> Question 18 is another example of a comprehension question. It requires you to draw an inference.
> Question 19 is a type of critical analysis question. It asks you to understand the author's persuasive technique or strategies.

Tips and Explanations:

16. The correct answer is C. For questions asking you about the order of steps or events, you need to focus on the part of the selection where the particular step is mentioned. Here, we can see that the question is asking about the occurrence of pressure on the tectonic plates. So, we need to focus on the last two sentences of the first paragraph: "The two plates will eventually shift or separate because the pressure on them is constantly increasing, and this build-up of energy needs to be released. When the plates shift or separate, we have an occurrence of an earthquake, also known as a seismic event." You also need to pay attention to the words in the question that are indicating the sequencing, such as "before," "after," "next," or "during." The question asks us: "What happens immediately *after* the pressure on the tectonic plates has become too great?" The word "after" shows that we need to determine the next step. The selection indicates that after the pressure builds up, it needs to be released. The passage states that the release of energy in this way causes an earthquake, which is also called "a seismic event." So, answer C is correct.

17. The correct answer is B. In order to arrange ideas within a selection, pay special attention to the sentences before and after the gap. The sentence before the gap and the sentence after the gap both talk about earthquake prediction systems, so the sentence that goes in the gap should also mention this topic. By process of elimination, we can exclude answers C, D, and E because these three answers mention earthquakes only in general, but not prediction systems in particular. Answer B is the best answer because it mentions saving lives, which relates to the practical use of the prediction systems, mentioned in the last sentence.

18. The correct answer is D. In order to draw an inference, you should make only a small logical step based on the information contained in the selection. Try to avoid making wild guesses. In this selection, we see that the sentence to be placed in the gap mentions that lives can be saved through prediction systems, while the last sentence states that "these prediction systems need to be more reliable in order to be of any practical use." Accordingly, we can surmise that at the time the selection was written, prediction systems were not reliable enough. Therefore, not as many lives were being saved as would have been possible if the systems had been more reliable.

19. The correct answer is A. For questions about the author's persuasive technique or strategies, you should focus on the sentences before the one that contains the final direct statement by the author. In the selection above, the author's final direct statement is that "these prediction systems need to be more reliable in order to be of any practical use." The author's final statement is dispassionate

because she precedes it with the scientific descriptions of earthquake waves and epicenters. [Note that "dispassionate" means objective or stated without strong emotion.]

Read the passage below and answer the two questions that follow.

In December 406 AD, departing from what is now called Germany, 15,000 warriors crossed the frozen Rhine River and traveled into the Roman Empire of Gaul. A new historical epoch would soon be established in this former Roman Empire.

Even though this period has diminished in historical significance in comparison to more recent events, the demise of the Roman Empire in the fifth century was certainly unprecedented. Today, the collapse of the Roman Empire remains significant because it marks the commencement of what we now call the Middle Ages, the six subsequent centuries that followed the demise of Roman rule.

20. According to the passage, the Roman Empire:
    A. was established during the Middle Ages.
    B. is now referred to as Germany.
    C. gradually collapsed throughout the Middle Ages.
    D. fell into ruin from 406 to 499 AD.
    E. has become more historically significant in recent times.

21. Which of the following outlines best describes the organization of the topics addressed in paragraphs I and II?
    A. **I.** Invasion of Germany in the fifth century; **II.** A comparison to recent current events
    B. **I.** Background to the Middle Ages; **II.** The demise of the Roman Empire
    C. **I.** Crossing the Rhine River for battle; **II.** Historical significance of the collapse of the Roman Empire
    D. **I.** A new historical epoch for the Roman Empire; **II.** Why the Roman Empire collapsed
    E. **I.** The invasion of the Roman Empire of Gaul; **II.** The beginning of the Middle Ages

> Question 20 is the type of comprehension question that requires you to understand the relationship between general and specific ideas in a selection.
> Question 21 is the type of comprehension question that asks you to understand the organizational scheme of a selection.

Tips and Explanations:

20. The correct answer is D. Be sure to look at the relationship between the general idea and the specific points for questions like this one. From the selection, we can see that the general idea is the history of the Roman Empire and that the specific ideas are the dates of the events. From the first sentence, we know that the Roman Empire of Gaul was invaded in 406. In the second paragraph, we read that the Roman Empire fell into demise in the fifth century, which is the 100-year period ending in 499 AD. Therefore, we can conclude that the Roman Empire fell into ruin from 406 to 499 AD.

21. The correct answer is E. For questions on organizational scheme like this one, you have to be careful that the answer is not too specific. In other words, you should normally try to choose the most general answer, without choosing an overgeneralization. Paragraph one of the passage talks about warriors crossing the frozen Rhine River and traveling into the Roman Empire of Gaul. Therefore, paragraph one focuses on the invasion of the Roman Empire of Gaul. Paragraph two explains that the Roman Empire remains significant because it marks the commencement (or beginning) of what we now call the Middle Ages.

Read the passage below and answer the three questions that follow.

American Major League Baseball consisted of only a handful of teams when the National League was founded in 1876. Yet, baseball has grown in popularity <u>by leaps and bounds</u> over the years.

This growth in popularity resulted in increased ticket sales for games and bolstered the profits of its investors. The increased demand from the public, in turn, precipitated the formation of a new division, known as the American League, in 1901.

Additionally, new teams have been formed from time to time in accordance with regional demand. This was the case with the Colorado Rockies in Denver, Colorado, and the Tampa Bay Rays in Tampa Bay, Florida.

22. The main purpose of the passage is:
   A. to give examples of some popular American baseball teams.
   B. to provide step-by-step information about the process of forming new baseball teams.
   C. to trace historical developments relating to the popularity of baseball.
   D. to criticize Americans who depend on baseball for entertainment.
   E. to compare and contrast the American and National Baseball Leagues.

23. Which one of the following phrases is closest in meaning to <u>by leaps and bounds</u> as it is used in the above text?
   A. with unbelievable speed
   B. by exceeding the boundaries
   C. with sporadic movements
   D. contrary to public opinion
   E. surpassing all requirements

24. Which of the following assumptions has influenced the writer?
   A. The increase in ticket sales is a direct result of increased investment.
   B. Those who invest money in baseball teams make too much profit from their investment.
   C. New baseball teams are more popular than established teams.
   D. The popularity of American baseball will continue to increase steadily in the future.
   E. The formation of baseball teams is based on certain economic principles.

> Question 22 is another critical analysis question on the author's purpose. We have seen one of these types of questions previously at number 3 above.
> Question 23 is the type of comprehension question that requires understanding the relationship between meaning and context.
> Question 24 is a critical analysis and evaluation question on identifying the author's assumptions.

Tips and Explanations:

22. The correct answer is C. Remember that in order to determine the author's purpose, you need to pay special attention to the last sentence of the first paragraph of a selection. This is where the author normally puts his or her thesis statement. From the phrase "over the years" in the author's thesis statement, we know that the passage has a historical focus. The selection begins with the founding date of 1876, moves on to a major event in 1901, and finishes by talking about more recent developments. So, the main purpose of the passage is to trace historical developments relating to the popularity of baseball. You may be tempted to choose answer B. However, the passage is not systematic enough to be classified as a step-by-step description.

23. The correct answer is A. This question requires that you understand the relationship between meaning and context. For these types of questions, you need to look in the selection for phrases that are synonyms or antonyms to the phrase in the question. The phrase "by leaps and bounds" is an idiomatic expression which refers to something that happens very quickly or dramatically. This is in contrast to the phrase "only a handful of teams" from the first sentence of the selection.

24. The correct answer is E. In order to identify the author's assumptions, you should evaluate the phrases that the author uses to expand his or her assertions. In the selection above, we can see that the author mentions that baseball's popularity "bolstered the profits of its investors." Additionally, the last paragraph states that "new teams have been formed from time to time in accordance with regional demand." Profit and demand are two economic concepts, so we can conclude that the formation of baseball teams is based on certain economic principles.

Read the passage below and answer the three questions that follow.

Airline travel is generally considered to be an extremely safe mode of transportation. Indeed, statistics reveal that far fewer individuals are killed each year in airline accidents than in crashes involving automobiles. _____ this safety record, airlines deploy ever-increasingly strict standards governing the investigation of aircraft crashes. Information gleaned from the investigation of aircraft crashes is important _____ it is utilized in order to prevent such tragedies from occurring again in the future.

25. Which one of the following statements is not supported by information contained in the passage?
    A. Airline crash investigation standards have become more rigorous.
    B. It is safer to travel in an airplane than in a car, according to the statistics.
    C. Traveling by air is normally very safe.
    D. Data from airline accidents is used to make improvements to airline safety standards.
    E. The number of airline accidents has decreased in recent years.

26. Which of the words or phrases, if inserted in order into the blanks in the passage, would help the reader better understand the sequence of events?
    A. Besides; as a result
    B. In spite of; because
    C. In addition to; although
    D. Apart from; while
    E. Despite; due to

27. Who is the writer's audience?
    A. a group of young children
    B. college students attending a lecture
    C. the general public
    D. pilots on a training program
    E. key members of airline staff

> Question 25 is a critical analysis and evaluation question that requires you to compare and contrast the ideas stated in the question to those within the selection.
> Question 26 is another question on using key transition and linking words, like question 7 above.
> Question 27 is a question on the author's intended audience. We have seen this question type before at number 11 above.

Tips and Explanations:

25. The correct answer is E. For questions that require you to compare and contrast the ideas stated in the question to the ideas stated within the selection, you should use the process of elimination technique. Answer A is stated in the passage because the adjective "rigorous" from the answer is synonymous with the phrase "ever-increasingly strict standards" from sentence three of the passage. The information from answers B and C is stated in the second sentence of the passage. The information from answer D is stated in the last sentence of the passage. Answer E is not stated in the passage. The final sentence mentions the hope that accidents will decline in the future, but we do not know if this is in fact the case.

26. The correct answer is B. Remember that for questions on transition and linking words, you need to look carefully at the information that is provided in the selection after the gap. Determine whether an additional supporting idea is being stated or if the author is changing the subject. "In spite of" fits in the first gap because there is a shift in tone from the idea of safety to the idea of accident prevention. "Because" fits in the second gap since there is a cause-and-effect relationship between acquiring the information and preventing the accidents.

27. The correct answer is C. Remember that in order to identify the intended audience, you should evaluate the vocabulary that the passage uses. The information in the passage is pitched to an audience of general members of the public. The vocabulary used in the passage is too advanced for a group of young children, while it is not sufficiently academic for college students attending a lecture. The information is not technical enough for pilots on a training program or for key members of airline staff.

Read the passage below and answer the three questions that follow.

Clones have been used for centuries in the field of horticulture. For instance, florists have traditionally made clones of geraniums and other plants by modifying cuttings and re-planting them in fresh soil. As a result, cloning is considered acceptable and predictable in the realm of plants and flowers.

However, the rapid development of science and technology means that cloning processes could now be used on humans. _____ .
This fear stems from the ethical ramifications that will inevitably occur if cloning is extended to the human species.

28. From the information in this passage, it is reasonable to infer that:
 A. The subject of cloning has become somewhat controversial recently.
 B. Cloning has fallen out of favor with horticulturalists.
 C. In spite of certain misgivings, many people support human cloning.
 D. Technological advances have impeded the use of cloning.
 E. Cloning on human beings could be used for positive purposes.

29. Which sentence, if inserted into the blank line in paragraph 2, would be most suitable for the author's audience and purpose?
 A. Dolly the sheep was an early example of how cloning could be successfully used on mammals.
 B. But human cloning and horticultural cloning are worlds apart.
 C. Many people believe that the cloning of human beings has sinister undertones.
 D. That is why scientists need to have reservations in this area.
 E. Clones are not exactly like those that we see in the movies.

30. Between paragraphs 1 and 2, the writer's approach shifts from:
   A. cause to effect
   B. problem to solution
   C. explanation to example
   D. approbation to misgivings
   E. advantages to disadvantages

> Question 28 is the type of question that asks you to make an inference. We have seen this type of question previously at number 18 above.
> Question 29 is another question on arranging ideas within a selection, like question 17 above.
> Question 30 is a further type of question on organizational schemes of selections, like question 21 above.

Tips and Explanations:

28. The correct answer is A. You will recall that in order to draw an inference, you should make only a small logical step based on the information contained in the selection. In the selection above, we can understand that the topic is controversial because the first paragraph states that "cloning is considered acceptable and predictable in the realm of plants and flowers." However, the second paragraph qualifies this statement by countering that there is a "fear [that] stems from the ethical ramifications" of using the process on humans.

29. The correct answer is C. For questions on arranging ideas within a selection, you have to pay special attention to the sentences before and after the gap. Here, we need a sentence that will link the idea of using cloning on humans, which is mentioned the first sentence, to the idea of the fears surrounding human cloning, which is stated in the last sentence. Sentence C does this because it contains the synonymous phrases "the cloning of human beings" [i.e., the idea of using cloning on humans] and "sinister undertones" [i.e., the idea of the fears surrounding human cloning].

30. The correct answer is D. Remember that for questions on organizational scheme, you have to be careful that the answer is not too specific. "Approbation" means approval. The idea of approbation sums up the first paragraph because it states that cloning is approved for horticultural uses. "Misgivings" means fears or concerns, such as the fears about the ethical considerations of human cloning, which are considered in the second paragraph. You may be tempted to choose answer E. However, the ethical considerations are more than a mere disadvantage since they bring fear along with them.

Read the passage below and answer the four questions that follow.

Working in a run-down laboratory near Paris, Marie Curie worked around the clock to discover a radioactive element. When she finally captured her quarry in 1902, she named it "radium" after the Latin word meaning ray.

Madame Curie should certainly be an inspiration to scientists today. She had spent the day blending chemical compounds which could be used to destroy unhealthy cells in the body. As she was about to retire to bed that evening, she decided to return to her lab. There she found that the chemical compound had become crystalized in the bowls and was emitting the elusive light that she sought.

Inspired by the French scientist Henri Becquerel, Curie won the Nobel Prize for Chemistry in 1903. Upon winning the prize, she declared that the radioactive element would be used only to treat disease and would not be used for commercial profit.

Today radium provides an effective remedy for certain types of cancer. Radium, now used for a treatment called radiotherapy, works by inundating diseased cells with radioactive particles. Its success lies in the fact that it eradicates malignant cells without any lasting ill effects on the body.

31. Which of the following is the best meaning of the word <u>quarry</u> as it is used in this passage?
    A. a precious commodity
    B. an unknown catalyst
    C. an object that is sought
    D. a chemical compound
    E. a source that emits light

32. According to the information in the passage, why is radium treatment used as a cancer therapy?
    A. because it is cost effective
    B. because it destroys cancerous cells
    C. because it has no long-term effects
    D. because it emits a glowing light
    E. because it derives from a radioactive element

33. What is the most appropriate title of the passage?
    A. Madame Curie: An Inventive Chemist
    B. The Discoveries of Madame Curie
    C. The Use of Radium to Treat Cancer
    D. Madame Curie: A Brief Biography
    E. The Discovery and Use of Radium

34. Which of the following phrases or sentences from the passage expresses an opinion rather than a fact?
    A. Marie Curie worked around the clock to discover a radioactive element.
    B. Madame Curie should certainly be an inspiration to scientists today.
    C. She had spent the day blending chemical compounds which could be used to destroy unhealthy cells in the body.
    D. Upon winning the prize, she declared that the radioactive element would be used only to treat disease and would not be used for commercial profit.
    E. Today radium provides an effective remedy for certain types of cancer.

> Question 31 is a comprehension question that asks you to identify varying interpretations of a word.
> Question 32 is like question 10 above. It is asking you to identify details from the selection.
> Question 33 is asking you to find the title for the selection. This is another type of main idea question, like question 8 above.
> Question 34 is like number 13 above. It is asking you to discern facts from opinions.

Tips and Explanations:

31. The correct answer is C. For vocabulary questions like this one, you need to bear in mind that the word provided will have different interpretations, depending on its context. "Quarry" can mean a hole in which one digs for rock. Alternatively, "quarry" can refer to something that is hunted or pursued. Also remember that for vocabulary questions, you need to look for synonyms in the passage. In sentence one, we see the word "discover." In the last sentence of paragraph two, we see the phrase "the elusive light that she sought." Therefore, we can surmise that "quarry" is something one wants to discover or an object being sought.

32. The correct answer is B. Be careful. Questions like this will have distractor answers which will reiterate phrases from the passage, although these phrases do not answer the question. We know that answer B is correct because the final sentence of the passage states: "Its success lies in the fact that it eradicates malignant cells without any lasting ill effects on the body." You may be tempted to choose answer C. However, answer C is too general since radium has long-term positive effects [i.e., destroying malignant cells] without having any long-term negative effects.

33. The correct answer is E. For main idea questions, as well as for questions on selecting a title for a selection, you will need to choose an answer that is neither too general nor too specific. Answers A, B, and D are much too general since the passage does not focus on the entire life and work of Madame Curie. Answer C is too specific because cancer treatment is mentioned in only the last paragraph. Therefore, "The Discovery and Use of Radium" is the best title for the passage.

34. The correct answer is B. For "opinion vs. fact" questions like this, look for modal verbs (should, would, may, might) and superlative adjectives that express opinions (the best, the most, etc). The notion whether someone should be an inspiration to others is a matter of personal opinion, so B is the best answer. You may be tempted to choose answer E. However, the adjective "effective" is factually qualified by the use of the phrase "certain types."

Read the passage below and answer the three questions that follow.

In Southern Spain and France, Stone Age artists painted stunning drawings on the walls of caves nearly 30,000 years ago. Painting pictures of the animals upon which they relied for food, the artists worked by the faint light of lamps that were made of animal fat and twigs.

In addition to having to work in relative darkness, the artists had to endure great physical discomfort since the inner chambers of the caves were sometimes less than one meter in height. Thus, the artists were required to crouch or squat uncomfortably as they practiced their craft.

Their paints were mixed from natural elements such as yellow ochre, clay, calcium carbonate, and iron oxide. However, many other natural elements and minerals were not used. An analysis of the cave paintings reveals that the colors of the paints used by the artists ranged from light yellow to dark black.

The artists utilized ochre and manganese as engraving tools in order first to etch their outlines on the walls of the caves. Before removing their lamps and leaving their creations to dry, they painted the walls with brushes of animal hair or feathers. Archeologists have also discovered that ladders and scaffolding were used in higher areas of the caves.

35. What was the last step in the process of Stone Age cave drawings?
    A. The paintings were etched.
    B. The paint was applied.
    C. The lamps were removed.
    D. The artwork was left to dry.
    E. The scaffolding was erected.

36. Which of the following best expresses the attitude of the writer?
    A. It is surprising that the tools of Stone Age artists were similar to those that artists use today.
    B. It is amazing that Stone Age artists were able to paint such beautiful creations in spite of the extreme conditions they faced.
    C. The lack of light in the caves had an effect on their esthetic quality.

D. It is predictable and banal that Stone Age artists would paint pictures of animals.
E. The use of natural elements in paint was not an environmentally-friendly practice.

37. Which sentence is least relevant to the main idea of the passage?
   A. Thus, the artists were required to crouch or squat uncomfortably as they practiced their craft.
   B. Their paints were mixed from natural elements such as yellow ochre, clay, calcium carbonate, and iron oxide.
   C. However, many other natural elements and minerals were not used.
   D. An analysis of the cave paintings reveals that the colors of the paints used by the artists ranged from light yellow to dark black.
   E. The artists utilized ochre and manganese as engraving tools in order first to etch their outlines on the walls of the caves.

> Question 35 asks you to determine the correct order of events or steps in a process, like question 16 above.
> Question 36 is like question 12 above. It is an evaluation question on understanding the author's viewpoint.
> Question 36 asks you to determine the relevance of ideas to the selection, like question 14 above.

Tips and Explanations:

35. The correct answer is D. We need to have a look at the first and second sentences of the last paragraph, which state: "The artists utilized ochre and manganese as engraving tools in order first to etch their outlines on the walls of the caves. Before removing their lamps and leaving their creations to dry, they painted the walls with brushes of animal hair or feathers." Be sure to read sentences like this one very carefully. The etching is the first step. The application of the paint is the second step. Removing the lamps is the third step, while leaving the paint to dry is the final step.

36. The correct answer is B. The attitude of the writer is that it is amazing that Stone Age artists were able to paint such beautiful creations in spite of the extreme conditions they faced. For questions like this one, look for adjectives in the passage that give hints about the author's point of view. The phrase "stunning drawings" in paragraph one indicates the author's amazement.

37. The correct answer is C. The article focuses on the natural elements that were used in the process of creating the drawings. The passage is therefore not concerned with other natural elements that were not used.

Look at the table of contents below from an introductory textbook on computer science in order to answer the two questions that follow.

Acknowledgments	1
Introduction	3
Part 1: Early Computers	11
Mainframe Size	22
Processing Capacity	38
Peripherals	45
Limitations	
Part 2: State-of-the-Art Hardware	
Platform Differences	69
Monitors	82
Printers and Scanners	102
Other Peripherals	125
Part 3: New Developments in Software	
Operating Systems	137
Word Processing	149
Data Processing	161
Multimedia	180
Part 4: The World of Online Technology	
Internet Search Engines	202
Email Interfaces	231
RSS Feeds	254
Uploading and Downloading	273
Recommendations	304
Bibliography	325
Index	337

38. Which part of the book is most likely to address the care and maintenance of computer equipment?
    A. Introduction
    B. Part 1
    C. Part 2
    D. Part 3
    E. Part 4

39. Where can the reader find the names of the people and organizations that the author has thanked for their support during the writing of the book?
    A. Acknowledgments
    B. Introduction
    C. Recommendations
    D. Bibliography
    E. Index

> Question 38 and 39 are similar to questions 1 and 2 above. However, here we are asked to evaluate a table of contents, rather than an index.

Tips and Explanations:

38. The correct answer is C. The care and maintenance of computer equipment would be discussed in Part 2 since computer equipment is synonymous with hardware.

39. The correct answer is A. "Acknowledge" means to recognize as important or to thank, so the names of the people and organizations that the author has thanked for their support during the writing of the book would appear in the acknowledgments section.

Read the passage below and answer the four questions that follow.

The world's first public railway carried passengers, even though it was primarily designed to transport coal from inland mines to ports on the North Sea. Unveiled on September 27, 1825, the train had 32 open wagons and carried over 300 people.

The locomotive steam engine was powered by what was termed the steam-blast technique. _____ . In this way, the steam created a draft of air which followed after it, creating more power and speed for the engine.

The train had rimmed wheels which ran atop rails that were specially designed to give the carriages a faster and smoother ride. While the small carriages could hardly be termed commodious, the locomotive could accelerate to 15 miles per hour, a record-breaking speed at that time.

Subsequently, the inventor of the locomotive, George Stephenson, revolutionized his steam engine by adding 24 further pipes. Now containing 25 tubes instead of one, Stephenson's second "iron horse" was even faster and more powerful than his first creation.

40. Which of the following is the best meaning of the word commodious as it is used in this passage?
    A. small
    B. uncomfortable
    C. spacious
    D. speedy
    E. smooth

41. Which sentence, if inserted into the blank line in paragraph 2, would best fit into the logical development of the passage?
    A. The chimney of the locomotive redirected exhaust steam into the engine via a narrow pipe.
    B. This technique was quite innovative at that time, making Stephenson its pioneer.
    C. Previous engines had used different propulsion devices that were not as powerful.
    D. Most of the passengers were unaware of the technology behind Stephenson's invention.
    E. Because of the power of the engine, it was important to ensure that the passengers would have a smooth ride.

42. Why was the second locomotive that Stephenson invented an improvement on his first?
    A. because it ran more smoothly
    B. because it was more comfortable

C. because it could carry more passengers
   D. because it contained more pipes and tubes
   E. because it ran with greater force and speed

43. From the information contained in the passage, it seems reasonable to infer which of the following?
   A. Many passengers were frightened about traveling on Stephenson's new locomotive.
   B. George Stephenson's inventions laid the basic foundations for modern day public trains and railways.
   C. Profits in the coal industry increased after the invention of the locomotive.
   D. Stephenson should have been able to invent a locomotive that could run faster.
   E. Stephenson's second locomotive carried more passengers than his first one.

> Question 40 is another comprehension question on the meaning of unknown words.
> Question 41 is the type of question that requires to you to arrange ideas logically within a selection.
> Question 42 a critical analysis question on identifying details from the passage.
> Question 43 is an example of a comprehension question that requires you to draw an inference.

Tips and Explanations:

40. The correct answer is C. Remember that for vocabulary questions like this one, you will need to look for synonyms or antonyms of the word in question. The sentence begins "While the small carriages could hardly be termed commodious . . ." so we know that the word "commodious" is the opposite of small. Accordingly, "spacious" is the correct answer.

41. The correct answer is A. When placing a sentence into a gap in a paragraph, you need to look carefully at the sentences before and after the gap in order to discover what idea unites the paragraph. In this case, the idea of steam, specifically the movement of steam, is mentioned in both of these two sentences. Sentence A also talks about the movement of steam, so it is the best answer.

42. The correct answer is E. The last sentence of the paragraph states that "Stephenson's second 'iron horse' was even faster and more powerful than his first creation." In other words, we can conclude that the second locomotive was an improvement because it ran with greater force and speed that the first one did.

43. The correct answer is B. The passage describes how George Stephenson invented the steam locomotive and the world's first public railway. Such inventions lay the basic foundations, which can later be improved upon with advances in technology. So, George Stephenson's inventions laid the basic foundations for modern day public trains and railways.

Read the passage below and answer the three questions that follow.

Highly concentrated radioactive waste is lethal and can remain so for thousands of years. Accordingly, the disposal of this material remains an issue in most energy-producing countries around the world. In the United States, for example, liquid forms of radioactive waste are usually stored in stainless steel tanks. For extra protection, the tanks are double-walled and surrounded by a concrete covering that is one meter thick. This storage solution is also utilized the United Kingdom, in most cases.

The long-term problem lies in the fact that nuclear waste generates heat as radioactive atoms decay. This excess heat could ultimately result in a radioactive leak. Therefore, the liquid needs to be cooled by pumping cold water into coils inside the tanks. However, the tanks are only a temporary storage solution. The answer to the long-term storage of nuclear waste may be fusing the waste into glass cylinders that are stored deep underground.

44. How are the tanks which are used for storing radioactive waste protected against leaks?
    A. They are encased in concrete.
    B. They only contain waste in liquid form.
    C. They provide a place where radioactive atoms can decay.
    D. They are combined with cold water.
    E. They are fused into glass cylinders.

45. Which of the following outlines best describes the organization of the topics addressed in paragraphs I and II?
    A. I. Radioactive Waste in the US and UK; II. Storage Problems
    B. I. Current Storage Solutions for Radioactive Waste; II. Potential Problems and Long-Term Solutions
    C. I. Radioactive Waste: The Long-Term Risks; II. Looking for Potential Solutions
    D. I. The Threat of Radioactive Waste; II. The Creation of Glass Cylinders
    E. I. Stainless Steel Storage Tanks for Radioactive Waste; II. The Generation of Heat and Potential for Leaks

46. Which of the following assumptions has most influenced the writer?
    A. The threat of a radioactive leak is exaggerated by the public.
    B. The storage of radioactive waste in stainless steel tanks is extremely dangerous.
    C. The United Kingdom normally follows practices that the United States has adopted.
    D. The underground storage of glass cylinders containing radioactive waste is going to be a very risky procedure.
    E. A radioactive leak would have disastrous consequences around the globe.

> Question 44 is a critical analysis question on identifying details from the passage.
> Question 45 is the type of comprehension question that asks you to understand the organizational scheme of a selection.
> Question 46 a critical analysis and evaluation question on identifying the author's assumptions.

Tips and Explanations:

44. The correct answer is A. The tanks are protected against leaks because they are encased in concrete. The fourth sentence of paragraph one states: "For extra protection, the tanks are double-walled and surrounded by a concrete covering that is one meter thick."

45. The correct answer is B. The first two sentences of paragraph one introduce the idea of radioactive waste generally, before moving on to talk about how the waste is stored at the present time, so the best title for paragraph one is "Current Storage Solutions for Radioactive Waste." Paragraph two begins by discussing the problems with storing the waste in this way and ends by giving an overview of possible solutions to these problems, so "Potential Problems and Long-Term Solutions" is the best title for paragraph two.

46. The correct answer is E. The author implies that a radioactive leak would have dire consequences since he opens the passage with this sentence: "Highly concentrated radioactive waste is lethal and can remain so for thousands of years."

Read the passage below and answer the four questions that follow.

Best known for his process of pasteurization, or the <u>eradication</u> of germs in liquid substances, Louis Pasteur was also the father of the modern rabies vaccine. In December of 1880, a friend who was a veterinary surgeon gave Pasteur two rabid dogs for research purposes.

Victims of bites from rabid dogs normally showed no symptoms for three to twelve weeks. By then, however, the patient would be suffering from convulsions and delirium, and it would be too late to administer any remedy. Within days, the victim would be dead.

So-called treatments at that time consisted of burning the bitten area of skin with red-hot pokers or with carbolic acid. _____ . Pasteur devoted himself to discovering a more humane and effective method of treatment for the disease.

His tests on rabid dogs confirmed that the rabies germs were isolated in the saliva and nervous systems of the animals. After many weeks of tests and experiments, Pasteur at last cultivated a vaccine from a weakened form of the rabies virus itself.

47. Which sentence, if inserted into the blank line in paragraph 3, would be most appropriate for the author's audience and purpose?
    A. These "remedies" often resulted in fatal trauma to the patients.
    B. Carbolic acid was also a common cleaning agent at that time.
    C. Dog owners were often unaware that their pets were infected with rabies.
    D. As medical professionals, you are aware of the dangers of these types of "treatments."
    E. Researchers today continue to cultivate new strains of the rabies vaccine.

48. Which of the following is the best meaning of the word <u>eradication</u> as it is used in this passage?
    A. cleansing
    B. reduction
    C. destruction
    D. amelioration
    E. assuagement

49. What are the symptoms of rabies infection, if it is left untreated?
    A. reddening of the skin
    B. burning sensation of the skin
    C. seizures and anxiety
    D. muscular contractions and forgetfulness
    E. mental disturbances and physical tremors

50. What is the most appropriate title of this passage?
    A. Pasteurization and the Rabies Vaccine
    B. The Life and Work of Louis Pasteur
    C. Pasteur's Discovery of the Rabies Vaccine
    D. Experimental Research on Rabid Dogs
    E. Uses of the Modern Rabies Vaccine

> Question 47 is another type of question on the author's intended audience.
> Question 48 is another vocabulary question.
> Question 49 is another identifying details question.
> Question 50 is a further question on finding the title for a selection.

Tips and Explanations:

47. The correct answer is A. This paragraph is devoted to discussing how patients used to be treated for rabies. Sentence A is the only option that talks about patients, so it is the best answer.

48. The correct answer is C. The sentence states: "Best known for his process of pasteurization, or the eradication of germs in liquid substances, Louis Pasteur was also the father of the modern rabies vaccine." Common sense tells us that germs are something negative that we would want to get rid of permanently, so "destruction" is the best answer. "Amelioration" and "assuagement" both mean to lessen the intensity of something, not to get rid of it permanently, so they are not the best answers.

49. The correct answer is E. Paragraph two states: "Victims of bites from rabid dogs normally showed no symptoms for three to twelve weeks. By then, however, the patient would be suffering from convulsions and delirium, and it would be too late to administer any remedy." "Delirium" means mental disturbances and "convulsions" means physical tremors.

50. The correct answer is C. The passage talks mostly about how Pasteur discovered the rabies vaccine, so "Pasteur's Discovery of the Rabies Vaccine" is the best title. You may be tempted to choose answer A. However, pasteurization is mentioned only in passing. Answer B is an overgeneralization. Answers D and E are too specific.

CBEST Practice Reading Test 2

Look at the extract from an index below to answer the two questions that follow.

```
Inventions  138–139
       classification as  139
       originality of  138
Inventors  140–155
       definition  141
       Edison, Thomas Alba  142–146
              gramophone  144
              incandescent lamp  146
              kinetoscope  143
              phonograph  142
              transmitter  145
       Marconi, Guglielmo  147–152
              radar  147
              telegraph  148–152
       Singer, Isaac Merritt  153–154
              treadle sewing machine  154
              universal acknowledgement of  153
Inventors and patents  153–170
       categories  157
```

1. Which page gives information about what determines whether a person is called "an inventor"?
   A. 138
   B. 139
   C. 140
   D. 141
   E. 157

2. How is the section of the book on individual inventors organized?
   A. alphabetically by inventor's last name
   B. chronologically by date of invention
   C. from the most to least significant
   D. by nationality of inventor
   E. by categories of use

Read the passage below and answer the three questions that follow.

Cancer occurs when cells in the body begin to divide abnormally and form more cells without control or order. There are some factors which are known to increase the risk of cancer. Smoking is the largest single cause of death from cancer in the United States. _____ , poor food choices increase cancer risk. Indeed, research shows that there is a definite link between the consumption of high-fat food and cancer.

If a cell divides when it is not necessary, a large growth called a tumor can form. These tumors can usually be removed, and in many cases, they do not come back. _____ , in some cases the cancer from the original tumor spreads. The spread of cancer in this way is called metastasis.

There are some factors which are known to increase the risk of cancer. Smoking is the single largest cause of death from cancer in the United States. One-third of the deaths from cancer each year are related to smoking, making tobacco use the most preventable cause of death in this country.

3. Which of the words or phrases, if inserted in order into the blanks of the passage, would help the reader better understand the sequence of events?
   A. However; Yet
   B. Although; Therefore
   C. Also; Thus
   D. Moreover; Even though
   E. In addition; However

4. What inference can be drawn from this passage?
   A. A low-fat diet can reduce the risk of cancer.
   B. The consumption of high-fat food has increased in recent years.
   C. Most cancer sufferers have made poor food choices.
   D. Smoking always causes cells to divide abnormally.
   E. The number of people who smoke is bound to decrease in the future.

5. What is metastasis?
   A. the abnormal growth of organs
   B. when a tumor begins to grow in size
   C. the growth of cancer in a cell
   D. the growth of cancer inside a tumor
   E. the spread of cancer from a tumor

Read the passage below and answer the three questions that follow.

The theory of multiple intelligences (MI) is rapidly replacing the intelligence quotient, or IQ. The IQ, long considered the only valid way of measuring intelligence, has come under criticism recently because it inheres in many cultural biases. For this reason, there has been a movement away from the IQ test, which is now seen as an indication of a person's academic ability. On the other hand, the theory of multiple intelligences measures practical skills such as spatial, visual, and musical ability.

Howard Gardner, the researcher who designed the system of multiple intelligences, posits that while most people have one dominant type of intelligence, most of us have more than one type. Of course, that's why they are known as multiple intelligences. As we will see today, this theory has important implications for teaching and learning.

6. Which of the following groups of statements best summarizes the main topics addressed in each paragraph?
   A. I. Disadvantages of the IQ test; II. The work of Howard Gardner
   B. I. The rise of the theory of multiple intelligences; II. Further information on the theory of multiple intelligences
   C. I. Cultural biases of the IQ test; II. The plurality of multiple intelligences
   D. I. IQ testing and academic performance; II. Dominant types of multiple intelligences
   E. I. The various aspects of multiple intelligences; II. Multiple intelligences in teaching and learning

7. Which of the following is the best meaning of the word posits as it is used in this passage?
   A. says
   B. points out
   C. suggests
   D. postulates
   E. considers

8.   The information the writer conveys in this passage is addressed mainly to:
   A. licensed psychologists attending a conference.
   B. college students attending a psychology class.
   C. college students attending an education class.
   D. the general public.
   E. young children.

Read the passage below and answer the two questions that follow.

Around the world today, more than a billion people still do not have fresh, clean drinking water available on a daily basis. Hundreds of thousands of people in developing countries die needlessly every year because of the consumption of unclean, disease-ridden water. In brief, fresh water saves lives. However, what has been understood only recently is that in order to improve the global water supply, those who manage water supplies must evaluate in more detail how developed countries consume their available drinking water. Without this evaluation, an ever-increasing number of individuals will continue to die from water-related diseases.

9.   We can conclude from the information in this passage that:
   A. water-related disease will decline in the future.
   B. the majority of water-related deaths could be avoided.
   C. children are the most vulnerable to water-related disease and death.
   D. developing countries manage their water supplies better than developed countries.
   E. governments will intervene to manage the world's water supplies.

10.   Which of the following assumptions has influenced the writer?
   A. Developing countries are culpable for the pollution of their own drinking water.
   B. The provision of fresh drinking water is the most pressing problem in recent current events.
   C. The consumption of water in developed countries could serve as a model to other countries.
   D. People living in developing countries should know better than to consume polluted water.
   E. The political climate of developing countries impedes their ability to have fresh drinking water.

Read the passage below and answer the four questions that follow.

In his first mathematical formulation of gravity, published in 1687, Sir Isaac Newton posited that the same force that kept the moon from being propelled away from the earth also applied to gravity at the earth's surface. While this finding, termed the Law of Universal Gravitation, is said to have been occasioned by Newton's observation of the fall of an apple from a tree in the orchard at his home, in reality the idea did not come to the scientist in a flash of inspiration, but was developed slowly over time.

It is because of Newton's work that we currently understand the effect of gravity on the earth as a global system. As a result of Newton's investigation into the subject of gravity, we know today that geological features such as mountains and canyons can cause variances in the earth's gravitational force. Newton must also be acknowledged for the realization that the force of gravity becomes less robust as the distance from the equator diminishes, due to the rotation of the earth, as well as the declining mass and density of the planet from the equator to the poles.

11.   What is the author's main purpose?
   A. to analyze natural phenomena
   B. to reconcile various gravitational theories
   C. to identify a reservation which Newton experienced

D. to emphasize the significance of Newton's achievement
E. to give background about Newton's life

12. Which of the following phrases is closest in meaning to the phrase <u>a flash of inspiration</u> as it is used in the above text?
    A. in hindsight
    B. with trepidation
    C. all of a sudden
    D. with clarity
    E. little by little

13. What is the pattern of organization of the passage?
    A. cause and effect
    B. general to specific
    C. explanation and example
    D. historical background and current applications
    E. theoretical development and new innovations

14. Which of the following is a factor in the diminishment of the force of gravity when one is closer to the equator?
    A. Because the relative weight of the Earth is higher in this particular geographical location
    B. Because the Earth's gravitational force has changed positions
    C. Because one is further from geographical features such as mountains and canyons
    D. Because the way in which the Earth rotates is different near the equator
    E. Because the distance to the poles has decreased

Read the passage below and answer the two questions that follow.

In the Black Hills, four visages protrude from the side of a mountain. The faces are those of four pivotal United States' presidents: George Washington, Thomas Jefferson, Theodore Roosevelt, and Abraham Lincoln. Washington was chosen on the basis of being the first president. Jefferson was selected because he was instrumental in the writing of the American Declaration of Independence. Lincoln was selected on the basis of the <u>mettle</u> he demonstrated during the American Civil War and Roosevelt for his development of Square Deal policy, as well as being a proponent of the construction of the Panama Canal.

15. From this passage, it seems reasonable to infer that these four presidents were chosen because:
    A. of their outstanding courage.
    B. their faces would be esthetically sympathetic to the natural surroundings.
    C. they helped to improve the national economy.
    D. they were considered the most popular among members of the public.
    E. their work was considered crucial to the progress of the nation.

16. Which of the following is the best meaning of the word <u>mettle</u> as it is used in this passage?
    A. emotion
    B. courage
    C. persistence
    D. persuasion
    E. determination

Use the graph below to answer the question that follows.

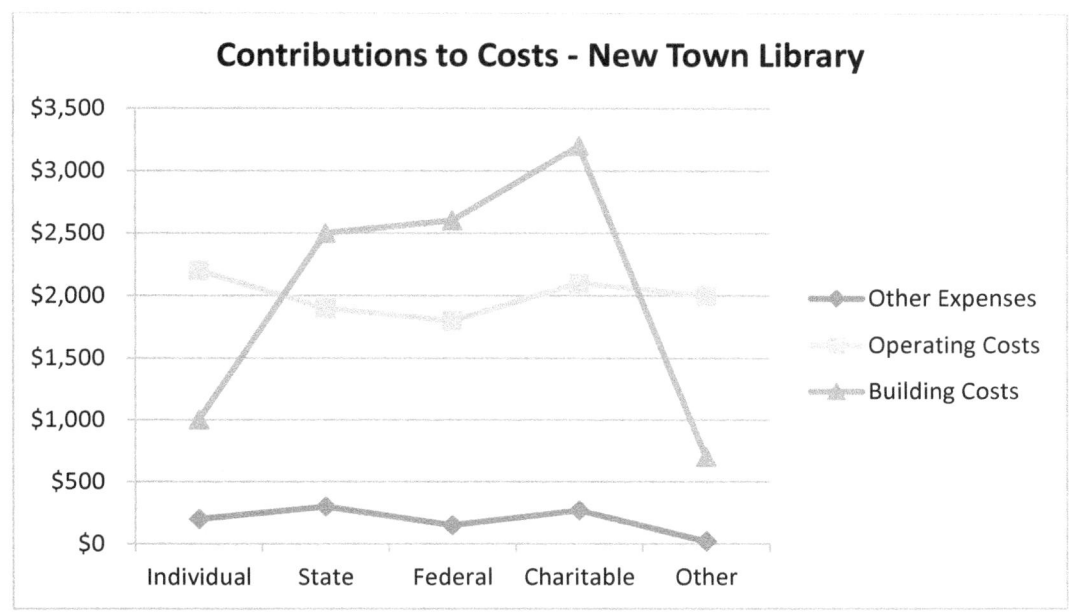

17. Which source provides the greatest amount of funding for the combined total of all three of the above types of costs?
    A. Private
    B. State
    C. Federal
    D. Charitable
    E. Other

Read the passage below and answer the three questions that follow.

[1]Socio-economic status, rather than intellectual ability, may be the key to a child's success later in life. [2]Consider two hypothetical elementary school students named John and Paul. [3]Both of these children work hard, pay attention in the classroom, and are respectful to their teachers. [4]Both boys have the same hobbies and musical tastes. [5]Nevertheless, Paul's father is a prosperous business tycoon, while John's has a menial job working in a factory.

[6]Despite the similarities in their academic aptitudes, the disparate economic situations of their parents means that Paul is nearly 30 times more likely than John to land a high-flying job by the time he reaches his fortieth year. [7]In fact, John has only a 12% chance of finding and maintaining a job that would earn him even a median-level income. [8]This outcome is inherently unfair because economic rewards should be judged by and distributed according to the worthiness of the employment to society as a whole, rather than according to social status or prestige.

18. What is the writer's primary persuasive technique?
    A. quoting from authorities
    B. appealing to emotion
    C. refuting opposing viewpoints
    D. predicting future consequences
    E. using statistical evidence

19. Which numbered sentence provides an opinion rather than a fact?
    A. Sentence 1
    B. Sentence 3
    C. Sentence 5
    D. Sentence 7
    E. Sentence 8

20. Which numbered sentence is least relevant to the main idea of the first paragraph?
    A. Sentence 1
    B. Sentence 2
    C. Sentence 3
    D. Sentence 4
    E. Sentence 5

Read the passage below and answer the three questions that follow.

The Hong Kong and Shanghai Bank Corporation (HSBC) skyscraper in Hong Kong is one of the world's most famous high-rise buildings. The building was designed so that it had many pre-built parts that were not constructed on site. This prefabrication made the project a truly international effort: The windows were manufactured in Austria, the exterior walls were fabricated in the United States, the toilets and air-conditioning were made in Japan, and many of the other components came from Germany.

The HSBC tower consists of 47 stories, which is an immense contrast to the twenty-story buildings in its vicinity. In fact, the previous buildings constructed on this site were limited by the soft and often water-logged ground in the surrounding area. _____ _____. This assessment was necessary in order to ensure that subsidence, and potential collapse of the new structure, could be averted.

21. Which sentence, if inserted into the blank line in paragraph two, would best fit into the logical development of the passage?
    A. Water-logging is a common problem in construction projects of this type.
    B. Many occupants of the neighboring buildings objected to the construction of the new skyscraper.
    C. For this reason, the groundwater supply had to be carefully assessed prior to construction of the HSBC building.
    D. Therefore, the new skyscraper was bound to dominate its structurally smaller neighbors.
    E. Subsidence is the phenomenon that occurs when a building shifts from its original position.

22. Which statement below best represents the writer's opinion?
    A. Prefabricated buildings are more international than those built on site.
    B. Countries should work together more often in construction projects.
    C. Careful planning is paramount for construction projects in urban settings.
    D. The HSBC building is well-known because many countries were involved in its construction.
    E. Construction projects should not disturb the natural groundwater supply.

23. Between paragraphs 1 and 2, the writer's approach shifts from:
    A. scientific detail to current problems
    B. interesting facts to potential problems
    C. analytical background to public inquiry
    D. explanation to example
    E. cause to effect

Read the passage below and answer the two questions that follow.

The study of philosophy usually deals with two key problem areas: human choice and human thought. A consideration of these problem areas is not an aspect of psychology or art. The first problem area, human choice, asks whether human beings can really make decisions that can change their futures. Conversely, it also investigates to what extent the individual's future is fixed and pre-determined by cosmic forces outside the control of human beings. In the second problem area, human thought, epistemology is considered. "Epistemology" means the study of knowledge; it should not be confused with ontology, the study of being or existence.

24. The primary purpose of the passage is:
    A. to compare two areas of an academic discipline.
    B. to explain key aspects of the work of a particular philosopher.
    C. to contrast psychological and artistic views on a particular topic.
    D. to investigate two troublesome aspects of human behavior.
    E. to provide historical background on the subject of philosophy.

25. Which sentence does not fit with the logical flow of the paragraph?
    A. A consideration of these problem areas is not an aspect of psychology or art.
    B. The first problem area, human choice, asks whether human beings can really make decisions that can change their futures.
    C. Conversely, it also investigates to what extent the individual's future is fixed and pre-determined by cosmic forces outside the control of human beings.
    D. In the second problem area, human thought, epistemology is considered.
    E. "Epistemology" means the study of knowledge; it should not be confused with ontology, the study of being or existence.

Read the passage below and answer the four questions that follow.

In the fall of 1859, a discouraged man was sitting in his run-down law office in Springfield, Illinois. He was fifty years old, in debt, and had been a lawyer for twenty years, earning on average 3,000 dollars a year. This man would later go on to do great things for his country. His name was Abraham Lincoln.

_____ these obvious financial constraints, some of Abraham Lincoln's associates had already begun to put forward the idea that he should run for president of the United States. Lincoln began to write influential Republican Party leaders for their assistance. By 1860, Lincoln had <u>garnered</u> more public support, after having delivered public lectures and political speeches in various states. _____ he was the underdog, Lincoln won 354 of the 466 total nominations at the Republican National Convention and was later elected President of the United States.

26. Which of the words or phrases, if inserted in order into the blanks of the passage, would help the reader better understand the sequence of events?
    A. Despite; As a result of
    B. Although; In spite of
    C. With; Even though
    D. During; Being
    E. In spite of; Although

27. Which of the following outlines best describes the organization of the topics addressed in paragraphs I and II?
    A. I. Lincoln's profession as a lawyer; II. Why Lincoln was the underdog
    B. I. Lincoln's life in Springfield; II. Lincoln's speeches and conventions
    C. I. Lincoln's biographical information; II. Lincoln's campaign and election

D. **I.** Lincoln's financial problems; **II.** Lincoln's campaign advisors
E. **I.** Lincoln's future achievement; **II.** How Lincoln won the election

28. Which of the following is the best meaning of the word <u>garnered</u> as it is used in this passage?
    A. taken
    B. earned
    C. forced
    D. achieved
    E. financed

29. Which of the following assumptions has influenced the writer?
    A. Successful politicians often encounter financial problems.
    B. A career in law provides a good background for a career in politics.
    C. Lincoln was an unlikely presidential candidate.
    D. Lincoln had an enduring legacy on the history of the United States.
    E. In order to win an election, a candidate must first obtain the support of his or her party.

Read the passage below and answer the three questions that follow.

The use of computers in the stock market helps to control national and international finance. These controls were originally designed in order to create long-term monetary stability and protect shareholders from catastrophic losses. Nevertheless, because of the high level of automation now involved in buying and selling shares, computer-to-computer trading could result in a downturn in the stock market.

Such a slump in the market, if not properly regulated, could bring about a computer-led stock market crash. _____ . For this reason, regulations have been put in place by NASDAQ, AMEX, and FTSE.

30. Which sentence, if inserted into the blank line in the passage, would best fit with the author's audience and purpose?
    A. Bonds and pension plans are also secure long-term investments.
    B. Trading shares via the internet has certainly increased nowadays.
    C. So be sure you frequently change the password you use for trading online.
    D. Needless to say, such an economic collapse would have disastrous consequences for the entire nation.
    E. The computer is an efficient tool for stock brokers as well, despite the risks.

31. Which sentence from the passage best expresses its central idea?
    A. The use of computers in the stock market helps to control national and international finance.
    B. These controls were originally designed in order to create long-term monetary stability and protect shareholders from catastrophic losses.
    C. Nevertheless, because of the high level of automation now involved in buying and selling shares, computer-to-computer trading could result in a downturn in the stock market.
    D. Such a slump in the market, if not properly regulated, could bring about a computer-led stock market crash.
    E. For this reason, regulations have been put in place by NASDAQ, AMEX, and FTSE.

32. Based on the information contained in the passage, what is a likely conclusion regarding computer-to-computer trading?
    A. Therefore, computer-to-computer trading is usually regarded as being safer now than it has been in the past.
    B. The volume of computer-to-computer trading is likely to decrease in the future because of the controls that have been introduced.
    C. Accordingly, the government needs to act now in order to introduce further controls on computer-to-computer trading.
    D. Hence, the fees charged by stockbrokers are bound to increase.
    E. Personal trading of shares via the internet will become more popular in the future.

Read the passage below and answer the four questions that follow.

Good nutrition is essential for good health. A healthy diet can help a person to maintain a good body weight, promote mental wellbeing, and reduce the risk of disease. So, you might ask, what does healthy nutrition consist of? Well, first of all, a healthy diet should include food from all of the major food groups. These food groups are carbohydrates, fruit, vegetables, dairy products, meat and other proteins, and fats and oils.

Besides this, it is also important to try to avoid processed or convenience food. Packaged food often contains chemicals, such as additives to enhance the color of the food or preservatives that give the food a longer life. Food additives are <u>deleterious</u> to health for a number of reasons. First of all, they may be linked to disease in the long term. In addition, they may block the body's ability to absorb the essential vitamins and minerals from food that are required for healthy bodily function.

33. What is the most appropriate title for this passage?
    A. Good Health and Wellbeing
    B. How to Eat a Balanced Diet
    C. The Dangers of Food Additives
    D. The Risks of Food Preservatives
    E. The Basics of Healthy Nutrition

34. Who is most likely to be the audience of this passage?
    A. Adults listening to a radio program on nutrition
    B. Medical doctors attending a seminar
    C. Participants in a weight-loss support group
    D. College students in a biology lecture
    E. Young children in an elementary school assembly

35. Which of the following words is closest in meaning to the word <u>deleterious</u> as it is used in the passage?
    A. insipid
    B. harmful
    C. impeding
    D. provoking
    E. preventative

36. According to the passage, what is the primary reason why manufacturers of processed food use additives?
    A. to make food more convenient
    B. to improve the appearance of the food
    C. to prevent the food from spoiling quickly

D. to add nutrients to the food
E. to remove harmful chemicals from the food

Read the passage below and answer the two questions that follow.

Although there are many different types and sizes of coins in various countries, vending machines around the world operate on the same basic principles.

The first check is the slot: coins that are bent or too large will not go in. Once inside the machine, coins fall into a cradle which weighs them. If a coin is too light, it is rejected and returned to the customer.

Coins that pass the weight test are then passed along a runway beside a magnet. Electricity passes through the magnet, causing the coin to slow down in some cases. If the coin begins to slow down, its metallurgic composition has been deemed to be correct.

The coin's slow speed causes it to miss the next obstacle, the deflector. Instead, the coin falls into the "accept" channel and the customer receives the product.

37. Based on the information in the passage, how is the metallurgical composition of a coin determined to be correct?
    A. By its weight
    B. By its increased velocity in the runway
    C. By whether it runs alongside the magnet
    D. By the electricity that has passed through the magnet
    E. By missing the deflector

38. The last step in testing the coin is:
    A. the slot
    B. determination of metallurgic composition
    C. the accept channel
    D. the deflector
    E. the customer's receipt of the product

Look at the table of contents below from an astronomy textbook in order to answer the two questions that follow.

Part 1–The Moon & Stars	1
The Phases of the Moon	24
Characteristics of Stars	38
The Constellations	49
Part 2–Comets & Meteors	
Cosmic Debris	57
Falling Stars	84
Characteristics of Comets and Meteors	120
Part 3–The Planets	
The Planets and Their Moons	138
Characteristics of the Planets	156
Planets and Gases	173
Life on the Planets	201
Part 4–Space Exploration	
Apollo: The Early Years	225
The Race to Space	251
After the Cold War	282
The Universe Beyond	311
List of Terms	333
Bibliography	341
Index	352

39. The reader wants to see if this textbook has made any reference to another book entitled: *The Earth and the Sky*. Where can the reader look to determine this the most quickly?
    A. Part 1
    B. Part 2
    C. List of Terms
    D. Bibliography
    E. Index

40. Which part of the book is most likely to discuss whether inter-planetary space travel will be undertaken in the future?
    A. Part 1
    B. Part 2
    C. Part 3
    D. Part 4
    E. Index

Read the passage below and answer the three questions that follow.

Michelangelo began work on the painting of the ceiling of the Sistine Chapel in the summer of 1508, assisted by six others who helped to mix his paint and plaster. However, as work proceeded, the artist dismissed each of his assistants one by one, claiming that they lacked the competence necessary to do the task at hand.

Described as the lonely genius, the painter himself often felt incompetent to complete the project entrusted to him by Pope Julius II. Having trained as a sculptor, Michelangelo had an extremely low opinion of his own painting skills. Yet, he went on to paint one of the most beautiful works in art history.

In spite of his frequent personal misgivings, he persevered to paint the ceiling with his vision of the creation of the universe. _____ . The scenes include the Separation of Light from Darkness, the Drunkenness of Noah, the Ancestors of Christ, and the Salvation of Mankind.

41. Which sentence below, if inserted into the blank in the last paragraph, would be most consistent with the logical flow of the passage?
    A. The nine scenes that he created ran in a straight line along the ceiling.
    B. He was originally commissioned to paint portraits of the twelve apostles.
    C. The Pope also had some misgivings about Michelangelo.
    D. People in the Vatican had grown accustomed to seeing the painter looking tired and disheveled.
    E. Michelangelo preferred to work alone and without distraction.

42. Why did Michelangelo dismiss his assistants?
    A. Because he decided that he preferred to mix his plaster by himself.
    B. Because their dismissal was requested by the Pope.
    C. Because he believed that they were inept craftsmen.
    D. Because he felt incompetent about his own abilities.
    E. Because they had no training in sculpture.

43. Which of the sentences from the passage, repeated below, expresses an opinion of the author rather than a fact?
    A. However, as work proceeded, the artist dismissed each of his assistants one by one, claiming that they lacked the competence necessary to do the task at hand.
    B. Described as the lonely genius, the painter himself often felt incompetent to complete the project entrusted to him by Pope Julius II.
    C. Having trained as a sculptor, Michelangelo had an extremely low opinion of his own painting skills.
    D. Yet, he went on to paint one of the most beautiful works in art history.
    E. In spite of his frequent personal misgivings, he persevered to paint the ceiling with his vision of the creation of the universe.

Read the passage below and answer the four questions that follow.

The pyramids at Giza in Egypt are still among the world's largest structures, even today. The monuments were constructed well before the wheel was invented, and it is notable that the Egyptians had only the most primitive, handmade tools to complete the massive project.

Copper saws were used to cut softer stones, as well as the large wooden posts that levered the stone blocks into their final places. Wooden mallets were used to drive flint wedges into rocks in order to split them. An instrument called an adze, which was similar to what we know today as a wood plane, was employed to give wooden objects the correct finish.

The Egyptians also utilized drills that were fashioned from wood and twine. In order to ensure that the stones were level, wooden rods were joined by strips of twine to check that the surfaces of the stone blocks were flat. Finally, the stone blocks were put onto wooden rockers so that they could more easily be placed into their correct positions on the pyramid.

44. The two tools which were used to place the stones into their final positions on the pyramid were made from which substance?
    A. flint
    B. copper
    C. twine
    D. stone
    E. wood

45. Between paragraphs 1 and 2, the writer's approach shifts from:
    A. scientific explanation to technical analysis
    B. reasoned argument to impassioned persuasion
    C. background information to specific details
    D. personal opinion to justification
    E. cause to effect

46. What is the writer's main purpose?
    A. to give a step-by-step explanation of the construction of the Giza pyramids
    B. to compare the construction of the Giza pyramids to that of modern day structures
    C. to give an overview of some of the main implements that were used to construct the Giza pyramids
    D. to highlight the importance of the achievement of the construction of the Giza pyramids
    E. to bring to light a misconception in previous accounts of the construction of the Giza pyramids

47. Which of the following assumptions has most influenced the writer?
    A. It is incredible that the Egyptians were able to construct the pyramids using only hand-made tools.
    B. It is a pity that the wheel was not available to the Egyptians during the construction of the pyramids at Giza.
    C. Modern construction projects could learn from the example of the Giza pyramids.
    D. The most difficult aspect of the project was placing the stones in the correct position on the pyramid.
    E. The pyramids could have been larger if more modern tools had been available.

Read the passage below and answer the three questions that follow.

She sighed in despair as he again showed no capacity to change his ways. "Why can't he see reason?" she wondered silently to herself for the umpteenth time that day.

The baby whimpered futilely in the next room, amid piles of unworn and unloved clothing. He too had learned that no matter how fiercely he cried, no one would come to his aid.

At times she felt like challenging her husband more strongly, or at least asking for an explanation of his behavior. Sadly, she too had been conditioned to learn that such actions were inutile. So, mute and hopeless, she stoically faced another day of domestic misery.

48. Which sentence from the passage best expresses its central idea?
    A. She sighed in despair as he again showed no capacity to change his ways.
    B. The baby whimpered futilely in the next room, amid piles of unworn and unloved clothing.
    C. He too had learned that no matter how fiercely he cried, no one would come to his aid.
    D. At times she felt like challenging her husband more strongly, or at least asking for an explanation of his behavior.
    E. Sadly, she too had been conditioned to learn that such actions were useless.

49. What is the best meaning of the word inutile as it is used in the passage?
    A. helpless
    B. useless
    C. impertinent
    D. desperate
    E. unimportant

50. Which word below best describes the woman's behavior to her spouse?
    A. amiable
    B. accustomed
    C. relentless
    D. futile
    E. resigned

ANSWER KEY AND EXPLANATIONS

Practice Reading Test 2

1. The correct answer is D. On line 5 of the index, you will see that page 141 gives the definition of the word "inventor."

2. The correct answer is A. The index mentions Edison, Marconi, and Singer in that order, so they are organized alphabetically by last name.

3. The correct answer is E. The gap in paragraph 2 introduces a second factor, so the phrase "in addition" is suitable. "However" is best for the second gap because there is a change in point of view, shifting from the idea that tumors can be removed to the idea that tumors sometimes spread.

4. The correct answer is A. The inference that a low-fat diet can reduce the risk of cancer can be drawn from this passage. The reverse of this idea is provided in the last sentence of the first paragraph, which states that "research shows that there is a definite link between the consumption of high-fat food and cancer."

5. The correct answer is E. Metastasis is the spread of cancer from a tumor. This idea is supported by the last two sentences of the second paragraph: "in some cases the cancer from the original tumor spreads. The spread of cancer in this way is called metastasis."

6. The correct answer is B. Paragraph 1 discusses the increase in popularity of the theory of multiple intelligences, while paragraph 2 gives further information on the theory of multiple intelligences, namely, some background information and a discussion of the educational implications. Therefore, answer B is the best because it is gives the general ideas of each paragraph. The other answers give only specific ideas from each of the paragraphs.

7. The correct answer is D. The words "posit" and "postulate" describe how theories are formed.

8. The correct answer is C. The last sentence talks about implications for teaching and learning, so the talk is being given in an education class.

9. The correct answer is B. We can conclude from the information in this passage that the majority of water-related deaths could be avoided. This idea is supported by the phrase "die needlessly" in the second sentence of the passage.

10. The correct answer is C. The assumption that has influenced the writer is that the consumption of water in developed countries could serve as a model to other countries. This is idea supported by the second to the last sentence in the passage, which mentions evaluating water management in developed countries: "what has been understood only recently is that in order to improve the global water supply, those who manage water supplies must evaluate in more detail how developed countries consume their available drinking water."

11. The correct answer is D. The author's main purpose is to emphasize the significance of Newton's achievement. This is supported by paragraph 2 sentence 1: "It is because of Newton's work that we currently understand the effect of gravity on the earth as a global system."

12. The correct answer is C. The phrase "flash of inspiration" means all of a sudden. The text gives the opposite idea at the end of the first paragraph, when it states that Newton's theories "developed slowly over time."

13. The correct answer is D. The pattern of organization of the passage is historical background in paragraph 1 and current applications in paragraph 2. The passage begins by talking about the year 1687, and then signals that it is moving to current applications by using the word "currently" in the first sentence of the second paragraph.

14. The correct answer is A. Gravity diminishes when one is closer to the equator because the relative weight of the Earth is higher in this particular geographical location. See the last sentence of the passage, which explains how gravity becomes less robust, in other words weaker, as the distance from the equator diminishes.

15. The correct answer is E. Each leader was considered crucial due to his unique contribution. Each of these different reasons is stated in the passage: "Washington was chosen on the basis of being the first president. Jefferson was selected because he was instrumental in the writing of the American Declaration of Independence. Lincoln was selected on the basis of the mettle he demonstrated during the American Civil War and Roosevelt for his development of Square Deal policy, as well as being a proponent of the construction of the Panama Canal."

16. The correct answer is B. In the context of war, we can surmise that "mettle" means courage.

17. The correct answer is D. You need to add up the three points, represented by the diamond, square, and triangle on each line, for each contributor (Private, State, Federal, Charitable, and Other).

Here are the totals for the five contributors:
Private – 3,400
State – 3,800
Federal – 4,550
Charitable – 5,750
Other – 2,790
Therefore, the category of charitable is the highest.

18. The correct answer is E. The writer's primary persuasive technique is using statistical evidence. Examples of this can be seen in the phrases "30 times more likely" and "12% chance" in paragraph 2.

19. The correct answer is E. Sentence 8 states: "This outcome is inherently unfair because economic rewards should be judged by and distributed according to the worthiness of the employment to society as a whole, rather than according to social status or prestige." The words "unfair," "should," and "worthiness" in this sentence demonstrate that an opinion is being given.

20. The correct answer is D. Sentence 4 states: "Both boys have the same hobbies and musical tastes." The hobbies and tastes of the boys are not related to their academic performance, which is the subject of this passage.

21. The correct answer is C. The sentence before the gap states: "the previous buildings constructed on this site were limited by the soft and often water-logged ground in the surrounding area." The sentence after the gap is: "This assessment was necessary in order to ensure that subsidence, and potential collapse of the new structure, could be averted." The phrase "this assessment" in the last sentence relates to the idea of carefully assessing the water supply, which is mentioned in answer C, as well as in the sentence preceding the gap.

22. The correct answer is C. The statement that best represents the writer's opinion is that "Careful planning is paramount for construction projects in urban settings." Paragraph 2 of the passage talks about how careful planning was needed because of the groundwater problem. Be careful if you wanted to choose answer D. This is a fact from the passage, not an opinion.

23. The correct answer is B. Between paragraphs 1 and 2, the writer's approach shifts from interesting facts to potential problems. The interesting facts in paragraph 1 are the various contributions to the project from different countries. The potential problems in paragraph 2 relate to the groundwater issue.

24. The correct answer is A. The primary purpose of the passage is to compare two areas of an academic discipline. Sentence 1 states: "The study of philosophy usually deals with two key problem areas: human choice and human thought." So, we can see that the academic discipline of philosophy is mentioned in this sentence. This sentence also talks about the two areas of the discipline, which are human choice and human thought.

25. The correct answer is A. The following sentence does not fit with the logical flow of the paragraph: "A consideration of these problem areas is not an aspect of psychology or art." The text is about philosophy, not psychology or art.

26. The correct answer is E. Both of these sentences begin with a shift in thought. In the first sentence, we shift from the idea that Lincoln had financial problems to the idea that he might become successful as US President. In the other sentence, we shift from the idea that Lincoln was the underdog to the idea that he won the election. The phrases "in spite of" and "although" show the kind of concession that is provided when shifts in thought like these occur in a passage.

27. The correct answer is C. Answer C is the best because it is gives the general ideas of each paragraph. Paragraph 1 talks about Lincoln's biographical information, while paragraph 2 describes his campaign and election. The other answers give only specific ideas from each of the paragraphs.

28. The correct answer is B. "Garnered" means earned. Lincoln earned his support though public speaking. You may have been tempted to choose answer D. However, the word "achieved" generally does not imply as much effort as does the word "garnered."

29. The correct answer is C. The assumption that influenced the writer is that Lincoln was an unlikely presidential candidate. The passage explains how Lincoln's age and financial situation were against him. Be careful if you wanted to choose answer D. The passage does not indicate or imply how Lincoln is regarded today, although you might be tempted to conclude this from your own knowledge.

30. The correct answer is D. The sentence before the gap states: "Such a slump in the market, if not properly regulated, could bring about a computer-led stock market crash." The sentence after the gap states: "For this reason, regulations have been put in place by NASDAQ, AMEX, and FTSE." The word "crash" from the previous sentence ties into the idea of "collapse" in answer D. The word "regulations" in the final sentence relates to how to control the possibility of such a crash or collapse.

31. The correct answer is C. This sentence from the passage best expresses its central idea: "Nevertheless, because of the high level of automation now involved in buying and selling shares, computer-to-computer trading could result in a downturn in the stock market." The passage goes on to talk about the consequences of improper regulations, so answer C is the best. You may have wanted to choose answer A, but it is not the best answer because the passage talks about how the controls are ineffective.

32. The correct answer is A. A likely conclusion regarding computer-to-computer trading is that it is usually regarded as being safer now than it has been in the past. Trading is safer now because of the regulations mentioned in paragraph 2. Answer B is a wild guess. Answers C, D, and E are not mentioned or implied in the passage.

33. The correct answer is E. The most appropriate title for this passage is "The Basics of Healthy Nutrition." This relates back to the rhetorical question in paragraph 1: "What does healthy nutrition consist of?"

34. The correct answer is A. The audience is most likely to be adults listening to a radio program on nutrition. The passage has a conversational tone, beginning sentences with words like "so" and "well."

35. The correct answer is B. We know that the word "deleterious" has a negative connotation because the passage is talking about disease at this point, so "harmful" is the best synonym.

36. The correct answer is B. According to the passage, the primary reason why manufacturers of processed food use additives is to improve the appearance of the food. In the second sentence of paragraph 2, we see that additives "enhance the color of food."

37. The correct answer is D. The metallurgical composition of a coin is determined to be correct by the electricity that has passed through the magnet. Paragraph 3 states: "Electricity passes through the magnet, causing the coin to slow down in some cases. If the coin begins to slow down, its metallurgic composition has been deemed to be correct." That is to say, the coin slows down because of the electricity that has passed through the magnet.

38. The correct answer is B. The last step in testing the coin is the determination of its metallurgic composition. This step is provided in the last sentence of paragraph 3: "If the coin begins to slow down, its metallurgic composition has been deemed to be correct." Be careful if you chose answer D. The deflector is not a step in the testing process, but rather an alternative outcome of the test.

39. The correct answer is D. If the reader wants to see if this textbook has made any reference to another book, he or she should look in the bibliography. A bibliography is a list of references to other materials.

40. The correct answer is D. Part 4 is most likely to discuss whether inter-planetary space travel will be undertaken in the future. This topic would be discussed in the chapter entitled "The Universe Beyond."

41. The correct answer is A. The mention of "nine scenes" in answer A relates to the phrase "the scenes" at the start of the next sentence.

42. The correct answer is C. Michelangelo dismissed his assistants because he believed that they were inept craftsmen. See the last sentence of paragraph 1, which states that "as work proceeded, the artist dismissed each of his assistants one by one, claiming that they lacked the competence necessary to do the task at hand."

43. The correct answer is D. The following sentence expresses an opinion of the author rather than a fact: "Yet, he went on to paint one of the most beautiful works in art history." The adjectival phrase "the most beautiful" indicates that an opinion is being given.

44. The correct answer is E. The two tools which were used to place the stones into their final positions on the pyramid were made from wood. Paragraph 3 mentions wooden rods and wooden rockers.

45. The correct answer is C. Between paragraphs 1 and 2, the writer's approach shifts from background information to specific details. Paragraph 1 describes tools in general, while paragraph 2 names specific tools.

46. The correct answer is C. The writer's main purpose is to give an overview of some of the main implements that were used to construct the Giza pyramids. The main purpose of the passage is implied in the last sentence of the first paragraph: "it is notable that the Egyptians had only the most primitive, handmade tools to complete the massive project.

47. The correct answer is A. The assumption that has most influenced the writer is that it is incredible that the Egyptians were able to construct the pyramids using only hand-made tools. The assumption that the outcome was incredible is shown by the contrast between the words "primitive" and "massive" in the last sentence of paragraph 1.

48. The correct answer is E. The following sentence from the passage best expresses its central idea: "Sadly, she too had been conditioned to learn that such actions were useless." The idea of conditioning is also mentioned in paragraph 2, with respect to the baby learning that crying would not change anything.

49. The correct answer is B. The woman sees that there is no point in such actions, so "useless" is the best answer.

50. The correct answer is E. The word "resigned" is a synonym of the phrase "mute and hopeless" in the last sentence.

CBEST Practice Reading Test 3

Look at the extract from an index below to answer the two questions that follow.

Exercise 238–297
   aerobic 243–254
   dangers of 274–286
      backache 280–282
      heart attack 277–279
      increased pulse 284
      muscle strain 276
      stiffness 274–275
   for fitness 255–263
      cycling 262
      football 260
      jogging 257
      swimming 255
      tennis 259
   history of 238–242
      17th century 238
      18th century 239
      19th century 240
      20th century 241
      21st century 242
   physical effects 264–273
      blood flow 265
      circulation 271–273
      heart rate 268–270
      muscles 267
   pregnancy and 287–297
      effect on mother's circulation 287–290
      effect on mother's heart 291–293
      seeking doctor's advice 295
Exercising your pet 125
Eye exercises 57

1. Where can the reader look to see whether the book contains information on how a person should warm up before undertaking aerobic exercise?
   A. Pages 243–254
   B. Pages 255–263
   C. Pages 264–273
   D. Pages 274–286
   E. Pages 287–297

2. How is the section of the book on the history of exercise organized?
   A. alphabetically
   B. chronologically
   C. by importance
   D. by type of exercise
   E. from least to most strenuous

Read the passage below and answer the three questions that follow.

Scientists have been conducting genetic engineering experiments for years. Gene splicing, the process whereby a small part of the DNA of one organism is removed and inserted into the DNA chain of another organism, has produced results like the super tomato. In order to create the super tomato, the gene resistant to cold temperatures on the DNA chain of a particular type of cold-water fish was isolated, removed, and inserted into an ordinary tomato plant. This resulted in a new type of tomato plant that can thrive in cold weather conditions.

However, gene splicing has become controversial lately. As animal rights groups have come more into prominence socially and politically, and people are more and more aware of the suffering of animals, many people question whether using animals in this way is medically reasonable, indeed whether it is even ethical or moral.

3. From this passage, it seems safe to conclude that
   A. the super tomato was the first case of gene splicing.
   B. the super tomato is only one example of gene splicing.
   C. DNA from tomatoes has also been inserted into certain types of fish.
   D. the interests of animal rights groups will soon fade from the public eye.
   E. most people object to gene splicing.

4. Which of the following statements gives the best summary of the main points of the lecture?
   A. Genetic engineering has a recent scientific background.
   B. DNA is the essential part of every living cell.
   C. The process of genetic engineering involves gene splicing of part of the DNA chain.
   D. The super-tomato can grow in severely cold conditions.
   E. Although gene splicing is not new, there have been ethical and moral debates about it recently.

5. Which of the following best describes the pattern of organization of this passage?
   A. problem and solution
   B. explanation and examples
   C. background and recent debates
   D. step-by-step instructions
   E. order of importance

Read the passage below and answer the four questions that follow.

[1]In 1804, Meriwether Lewis and William Clark began an expedition across the western United States. [2]This area was then known as the Louisiana Territory. [3]The two men had met years earlier and established a long-lasting friendship. [4]At that time, Lewis was well-known as possessing an outgoing and amiable personality. [5]When Lewis was later a young captain in the army, he received a letter from President Thomas Jefferson offering him funding to explore the Western country.

[6]With Jefferson's permission, Lewis offered a partnership in the expedition to his trusted friend Clark. [7]When their journey had safely concluded 8,000 miles later, President Jefferson purchased the

Louisiana Territory for fifteen million dollars. ⁸Thus, the most important land acquisition in the history of the United States took place.

6. Which numbered sentence provides an opinion rather than a fact?
   A. Sentence 3
   B. Sentence 4
   C. Sentence 6
   D. Sentence 7
   E. Sentence 8

7. Which numbered sentence is least relevant to the main idea of the first paragraph?
   A. Sentence 1
   B. Sentence 2
   C. Sentence 3
   D. Sentence 4
   E. Sentence 5

8. The main purpose of the passage is:
   A. to give the background to Lewis and Clark's westward expedition.
   B. to defend the purchase of the Louisiana Territory.
   C. to state a crucial decision made by Thomas Jefferson.
   D. to provide biographical information on Lewis and Clark.
   E. to compare the skills of Lewis and Clark.

9. Why did Clark travel with Lewis?
   A. Lewis needed someone to do menial jobs.
   B. Clark felt he needed a partner.
   C. Because of financial restrictions.
   D. Because they were good friends.
   E. Because President Jefferson requested it.

Read the passage below and answer the three questions that follow.

The Watergate burglary had many aspects, but at its center was President Richard Nixon. _____ the investigation of the burglary, government officials denied involvement in the crime. An extensive cover-up operation followed in an attempt to conceal those who were involved in planning the break-in. Yet, this <u>subterfuge</u> failed when the FBI investigated the one-hundred-dollar bills that were found in the pockets of the burglars. After making inquiries, the FBI discovered that this money originated from the Committee for the Re-election of the President, thereby confirming governmental involvement. _____ , individuals who had entered the highest branches of the American government to serve and protect the people went to prison instead.

10. Which of the words or phrases, if inserted in order into the blanks of the passage, would help the reader better understand the sequence of events?
    A. However; Yet
    B. Although; As a result
    C. While; Finally
    D. During; However
    E. Throughout; In the end

11. What is the main reason why the cover-up of the Watergate break-in failed?
    A. because the Committee for the Re-election of the President denied involvement
    B. because of the subterfuge of the FBI
    C. because the burglars' money was traced back to a governmental organization

D. because its ringleaders went to prison
E. because of the number of governmental officials in high-level positions

12. Which of the following is the best meaning of the word <u>subterfuge</u> as it is used in this passage?
    A. discovery
    B. concealment
    C. dishonesty
    D. investigation
    E. crime

Read the passage below and answer the three questions that follow.

Our ability to measure brain activity is owing to the research of two European scientists. It was in 1929 that electrical activity in the human brain was first discovered. Hans Berger, the German psychiatrist who made the discovery, was despondent to find out, however, that many other scientists quickly dismissed his research. The work of Berger was confirmed three years later when Edgar Adrian, a Briton, clearly demonstrated that the brain, like the heart, is profuse in its electrical activity. Because of Adrian's work, we know that the electrical impulses in the brain are a mixture of four different frequencies. _____ .
These four frequencies are called alpha, beta, delta, and theta.

13. Which sentence, if inserted into the blank line in the paragraph, would best fit into the logical development of the passage?
    A. By "frequency," we are referring to the number of electrical impulses that occur in the brain per second.
    B. Improper sleep patterns can cause brain frequencies to become irregular.
    C. Because of the stress of modern life, many people today suffer from interruptions to the natural electrical activity in their brains.
    D. Adrian was often called "the mad genius" by his contemporaries.
    E. Adrian's work soon began to outshine that of Berger.

14. The information the writer conveys in this passage is addressed mainly to:
    A. doctors attending a professional seminar.
    B. practicing brain surgeons.
    C. a television documentary on brain research.
    D. the general public.
    E. elementary schoolchildren.

15. What is the writer's opinion regarding the work of Hans Berger?
    A. It was proper that his work was dismissed by the scientific community.
    B. Berger's work was inferior to that of Adrian.
    C. Berger's work paved the way for the research of Adrian.
    D. Berger should have been more self-promoting about his discovery.
    E. Berger's work was one of the most important discoveries of the 20th century.

Read the passage below and answer the three questions that follow.

The most significant characteristic of any population is its age-sex structure, defined as the proportion of people of each gender in various age groups. The age-sex structure determines the potential for reproduction, and therefore population growth. Thus, the age-sex structure has social policy implications. For instance, a population with a high proportion of elderly citizens needs to consider its governmentally-funded pension schemes and health care systems. Conversely, a greater percentage of young children in the population might imply that educational funding and child welfare policies need to be evaluated. Accordingly, as the composition of a population changes over time, the government may need to re-evaluate its funding priorities.

16. Governmental funding decisions should primarily be based on:
    A. the composition of the age-sex groups within its population.
    B. the number of elderly citizens in its population.
    C. the percentage of children in its population.
    D. reproduction rates.
    E. social policy limitations.

17. What is the writer's primary persuasive technique?
    A. giving emotional pleas
    B. citing known authorities
    C. predicting opposing viewpoints
    D. listing priorities in order of importance
    E. using compelling examples

18. Which of the following assumptions has most greatly influenced the writer?
    A. Health care systems are one of the most important needs of society.
    B. The government needs to do more in order to support senior citizens.
    C. The number of young children in the population has risen at an alarming rate.
    D. Society needs to consider the requirements of all of its members and balance competing needs carefully.
    E. The conflicting interests of various social groups put an unnecessary strain on the government.

Use the graph below to answer the question that follows.

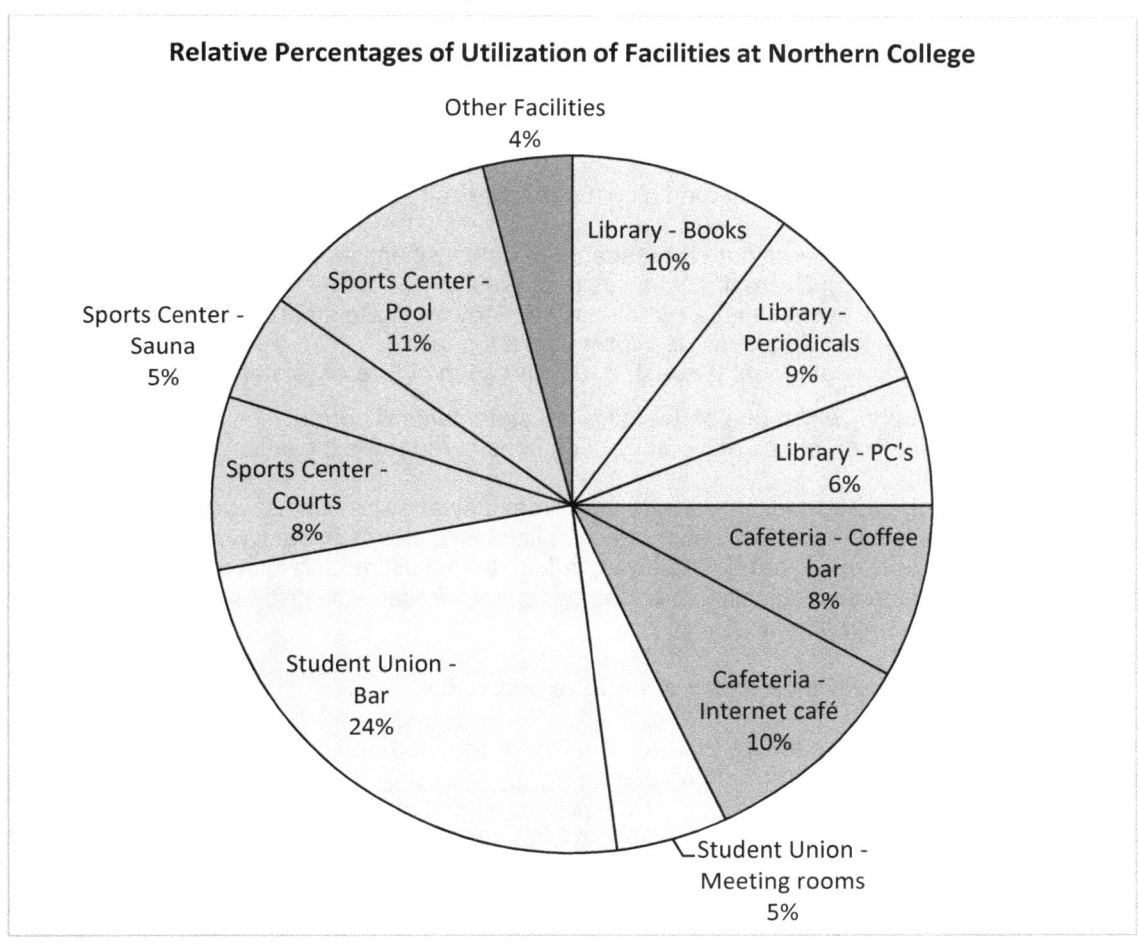

19. According to the graph, which area of the university has the highest percentage of use in total?
   A. Library
   B. Cafeteria
   C. Student Union
   D. Sports Center
   E. Other Facilities

Read the passage below and answer the three questions that follow.

I. In 1749, British surveyors spotted a high peak in the distant range of the Himalayas. More than 100 years later, in 1852, another survey was completed, which confirmed that this peak was the highest mountain in the world. Later named Mount Everest, this peak was unquestionably considered to be the world's highest mountain.

II. However, in 1986 George Wallerstein from the University of Washington posited that another Himalayan mountain, named K-2, was higher than Everest. It took an expedition of Italian scientists, who used a surfeit of technological devices, to disprove Wallerstein's claim.

20. What is the nearest synonym to the word surfeit in the passage?
    A. enhancement
    B. plethora
    C. collection
    D. survey
    E. variety

21. Which of the following groups of statements best summarizes the main topics addressed in each paragraph?
    A. **I.** Himalayan mountain expeditions; **II.** Research of Italian scientists
    B. **I.** The discovery of Mount Everest; **II.** University of Washington research
    C. **I.** Expeditions and discoveries in the Himalayas; **II.** Proving Wallerstein's claim
    D. **I.** British explorers in the Himalayas; **II.** Wallerstein's research
    E. **I.** Discovery and survey of Mount Everest; **II.** Confirmation of Everest's status

22. According to the passage, which one of the following statements is correct?
    A. Since 1749, Mount Everest has universally been considered to be the tallest mountain in the world.
    B. Wallerstein fell into disrepute in the academic community after his claims were disproved.
    C. The University of Washington fully supported Wallerstien's claims about K-2.
    D. The Italian team confirmed that Everest was, in fact, the tallest mountain in the world.
    E. In spite of a lack of technologically-advanced equipment, Italian scientists were able to refute Wallerstein's hypothesis.

Read the passage below and answer the four questions that follow.

Owing to the powerful and destructive nature of tornadoes, there are, perhaps not surprisingly, a number of myths and misconceptions surrounding them. _____
_____ . Yet, waterspouts, tornadoes that form over bodies of water, often move onshore and cause extensive damage to coastal areas. In addition, tornadoes can accompany hurricanes and tropical storms as they move to land. Another common myth about tornadoes is that damage to built structures, like houses and office buildings, can be avoided if windows are opened prior to the impact of the storm.

Drivers often attempt to outrun tornadoes in their cars, but it is extremely unsafe to do so. Automobiles offer very little protection when twisters strike, so drivers should abandon their vehicles and seek safe shelter. Mobile homes are extremely vulnerable, so residents of these homes should go to the underground floor of the sturdiest nearby building. In the case of a building having no underground area, a person should go to the lowest floor of the building and place him or herself under a piece of heavy furniture.

23. What inference about the public's knowledge of tornadoes can be drawn from the passage?
    A. A large number of people know how to avoid tornado damage.
    B. Most people appreciate the risk of death associated with tornadoes.
    C. Some members of the public know how to regulate the pressure inside buildings.
    D. Many people are not fully aware of certain key information about tornadoes.
    E. Many members of the public have an irrational fear of tornadoes.

24. Based on the information contained in the passage, which of the following best explains the term waterspouts in paragraph 1?
    A. Tornadoes that move away from coastal areas
    B. Tornadoes that occur over oceans, rivers, and lakes
    C. Tornadoes that occur onshore

D. Tornadoes that accompany tropical storms and hurricanes
E. Tornadoes that damage built structures

25. Which sentence if placed in the line in paragraph 1 would be the most consistent with the writer's purpose and audience?
    A. Indeed, the highest number of deaths and injuries are not caused by the winds themselves, but by flying objects and other debris.
    B. For instance, many people mistakenly believe that tornadoes never occur over rivers, lakes, and oceans.
    C. Therefore, public safety is of the utmost importance when a tornado strikes.
    D. For this reason, local governments must act quickly to put early severe weather warning systems into place.
    E. Tornadoes are often far different than those depicted in action movies.

26. According to the passage, what is the safest place to be when a tornado strikes?
    A. an abandoned vehicle
    B. mobile homes
    C. the basement of a building
    D. under a piece of sturdy furniture
    E. under a bridge

Read the passage below and answer the two questions that follow.

Jean Piaget was the most influential thinker in the area of child development in the twentieth century. Due to his training as a biologist, Piaget theorized that children go through a stage of assimilation as they grow to maturity. Assimilation refers to the process of transforming one's environment in order to bring about its conformance to innate or inborn cognitive processes. For instance, schemes used in infant breast feeding and bottle feeding are examples of assimilation. That is because the child utilizes his or her innate capacity for sucking to complete both tasks.

27. Which of the following sentences from the passage expresses an opinion rather than a fact?
    A. Jean Piaget was the most influential thinker in the area of child development in the twentieth century.
    B. Due to his training as a biologist, Piaget theorized that children go through a stage of assimilation as they grow to maturity.
    C. Assimilation refers to the process of transforming one's environment in order to bring about its conformance to innate or inborn cognitive processes.
    D. For instance, schemes used in infant breast feeding and bottle feeding are examples of assimilation.
    E. That is because the child utilizes his or her innate capacity for sucking to complete both tasks.

28. Which sentence from the passage best expresses its central idea?
    A. Jean Piaget was the most influential thinker in the area of child development in the twentieth century.
    B. Due to his training as a biologist, Piaget theorized that children go through a stage of assimilation as they grow to maturity.
    C. Assimilation refers to the process of transforming one's environment in order to bring about its conformance to innate or inborn cognitive processes.
    D. For instance, schemes used in infant breast feeding and bottle feeding are examples of assimilation.
    E. That is because the child utilizes his or her innate capacity for sucking to complete both tasks.

Read the passage below and answer the two questions that follow.

Inherent social and cultural biases pervaded the manner in which archeological findings were investigated during the early nineteenth century because little attention was paid to the roles that wealth, status, and nationality played in the recovery and interpretation of the artifacts. _____ , in the 1860s Charles Darwin established the theory that human beings are the ultimate product of a long biological evolutionary process. Darwinian theory infiltrated the discipline of archeology and heavily influenced the manner in which archeological artifacts are now recovered and analyzed. -_____ Darwinism, there has been a surge in artifacts excavated from Africa and Asia.

29. Which of the words or phrases, if inserted in order into the blanks of the passage, would help the reader better understand the sequence of events?
   A. Then; As a result of
   B. Although; In spite of
   C. While; Finally
   D. During; However
   E. Yet; In the end

30. Based on the information contained in the passage, what is a likely conclusion regarding archeological methods?
   A. They need to remain static to be useful.
   B. They should create cultural differences.
   C. They have developed a good deal when compared to earlier centuries.
   D. They should not have been rectified in countries in the Far East.
   E. They must be based on Darwinian theory in order to be valid.

Read the passage below and answer the three questions that follow.

The ancient Egyptians used eye shadow over 5,000 years ago. The cosmetic was used for personal beautification, as well as for practical reasons. Consisting of a paste made from malachite, a copper salt that was bright green, the eye paint protected against glare from the sun, in addition to being an attractive color. On her upper eye lids, Cleopatra wore blue eye shadow made of ground lapis lazuli stone, much like other women of her day.

The queen used green malachite as an accent below her eyes, and kohl, which consisted of lead sulfide, to provide color to her eyelashes and eyebrows. Red ochre, iron-based clay, provided her with lip and cheek color. Henna, a reddish-brown dye that was derived from a bush, was also commonly used by women in those days as a nail polish. _____
_____ . The use of this particular cosmetic was not limited to women. Men also used the substance to darken their hair and beards.

31. Which sentence, if inserted into the blank in paragraph 2, would be the most consistent with the author's purpose?
   A. Extracts from the henna bush could also be used for medicinal purposes.
   B. The henna was thickened with tannin from the bark or fruit of various trees in order to be suitable for cosmetic use.
   C. Henna is also used nowadays for decoration of the hands and feet.
   D. Certain environmental activists are concerned about the use of henna in this way.
   E. The henna plant has often been the subject of botanical investigations.

32. Which of the following outlines best describes the organization of the topics addressed in paragraphs I and II?
    A. I. Cosmetic uses of malachite; II. The beautification of Cleopatra
    B. I. Cosmetics in ancient Egypt; II. Cosmetic uses of henna
    C. I. The ancient Egyptians; II. The use of minerals in cosmetics
    D. I. Ancient Egyptian eye shadow; II. Other ancient cosmetics
    E. I. The history of personal beautification; II. Uses of cosmetics by men

33. What word best describes the style of writing in this passage?
    A. argumentative
    B. persuasive
    C. informative
    D. condemning
    E. balanced

Read the passage below and answer the three questions that follow.

Acid has been present in rain for millennia, naturally occurring from volcanoes and plankton. However, scientific research shows that the acid content of rain has increased dramatically over the past two hundred years, in spite of humanity's recent attempts to control the problem.

Rain consists of two elements, nitrogen and sulfur. When sulfur is burned, it transforms into sulfur dioxide. Nitrogen also oxides when burned. When released from factories into the atmosphere, both sulfur dioxide and nitrogen oxide react with the water molecules in rain to form sulfuric acid and nitric acid, respectively.

Factories and other enterprises have built high chimneys in an attempt to carry these gases away from urban areas. Nevertheless, the effect of the structures has been to spread the gases more thinly and widely in the atmosphere, thereby exacerbating the problem.

The acid in rain also emanates from automobile exhaust, domestic residences, and power stations. The latter have been the culprit of the bulk of the acid in rainwater in recent years. Since the pollutants are carried by the wind, countries can experience acid rain from pollution that was generated in countries thousands of miles away.

34. Which one of the following phrases is closest in meaning to the latter have been the culprit of the bulk as it is used in the above text?
    A. Automobile exhaust has caused the majority of acid rain.
    B. Automobile exhaust, domestic residences, and power stations have equally contributed to the creation of acid rain.
    C. Power stations are more widespread geographically than other causes of acid rain.
    D. Power stations generate a great deal of pollution that is carried by the wind.
    E. Power stations have been the largest contributor to the problem.

35. Between paragraphs 2 and 3, the writer's approach shifts from:
    A. scientific explanation to current problems
    B. chemical analysis to scientific inquiry
    C. historical background to current problems
    D. scientific inquiry to possible solutions
    E. cause to effect

36. Which detail from the passage best supports the writer's main idea?
   A. When sulfur is burned, it transforms into sulfur dioxide.
   B. When released from factories into the atmosphere, both sulfur dioxide and nitrogen oxide react with the water molecules in rain to form sulfuric acid and nitric acid, respectively.
   C. Nevertheless, the effect of the structures has been to spread the gases more thinly and widely in the atmosphere, thereby exacerbating the problem.
   D. The acid in rain also emanates from automobile exhaust, domestic residences, and power stations.
   E. Since the pollutants are carried by the wind, countries can experience acid rain from pollution that was generated in countries thousands of miles away.

Read the passage below and answer the two questions that follow.

The tradition of music in the western world originated in the genre of chanting. Chant, a monophonic form of music, was the dominant mode of music prior to the thirteenth century. The semantic origins of the word "monophonic" are interesting to etymologists. "Mono" is from a Greek word which means one thing alone or by itself. "Phonic" is also Greek in origin, and it means sound. Accordingly, monophonic music consists of only one sound or voice that combines various notes in a series. Polyphonic music appeared during the early Renaissance period. In contrast to monophonic music, polyphonic music consists of more than one voice or instrument, and it combines the notes from the different sources together simultaneously.

37. Which sentence is least relevant to the main idea of the passage?
   A. The tradition of music in the western world originated in the genre of chanting.
   B. Chant, a monophonic form of music, was the dominant mode of music prior to the thirteenth century.
   C. The semantic origins of the word "monophonic" are interesting to etymologists.
   D. "Mono" is from a Greek word which means one thing alone or by itself.
   E. Polyphonic music appeared during the early Renaissance period.

38. Who is this passage most likely addressed to?
   A. the general public
   B. a college class on music theory
   C. a student during a music lesson
   D. a group of classical composers
   E. elementary age children

Look at the table of contents below from a horticultural textbook in order to answer the two questions that follow.

INTRODUCTION	1
CHAPTER ONE: Trees and the Environment	
New Tree Growth	2
Photosynthesis	23
Use of Fertilizers	37
Greenhouse Effect	45
CHAPTER TWO: Conifers	
Pine	50
Spruce	89
Cedar	101
Fir	132
CHAPTER THREE: Deciduous Trees	
Oak	157
Maple	198
Walnut	206
Sycamore	234
CHAPTER FOUR: Fruit-bearing Trees	
Apple	268
Orange	273
Pear	281
Peach	299
GLOSSARY	315
BIBLIOGRAPHY	331
INDEX	348

39. Which part of the book is likely to contain information on why the leaves of certain trees change color in the autumn?
    A. Introduction
    B. Chapter One
    C. Chapter Two
    D. Chapter Three
    E. Chapter Four

40. A reader wants to find the definition of the phrase "environmental runoff." Where can the reader find this information the most quickly?
    A. Introduction
    B. Chapter One
    C. Glossary
    D. Bibliography
    E. Index

Read the passage below and answer the two questions that follow.

Baking a cake is easy, provided you have a good oven and the correct ingredients. For a moist and fluffy cake, you should first of all pre-heat the oven to 350 degrees Fahrenheit. Be absolutely sure that the oven is pre-heated to the correct temperature. While the oven is pre-heating, you can grease and flour your cake pan and mix your ingredients together.

Before adding the wet ingredients, mix the dry ingredients together. The latter consist of one and a half cups of sugar, one teaspoon of salt, two teaspoons of baking soda, and two cups of sifted flour, which should be mixed well in a large bowl. However, before proceeding with the mixture, ensure that the bowl is of a sufficient size to accommodate all of the ingredients. Now add one-half cup of vegetable shortening, two eggs, one cup of whole milk, and a teaspoon of vanilla.

Put the mixture into the cake pan, bake for 30 minutes, and enjoy!

41. Based on the instructions above, it is likely that failing to pre-heat the oven will result in:
    A. damage to the oven.
    B. the cake being burnt.
    C. the cake taking longer to bake.
    D. insufficient time to prepare the cake pan.
    E. the cake being dry and dense.

42. What should one do after preparing the cake pan?
    A. add the wet ingredients
    B. check that the mixing bowl is large enough
    C. mix the sugar, salt, baking soda, and flour together
    D. check the oven temperature
    E. add the shortening, eggs, milk, and vanilla

Read the passage below and answer the four questions that follow.

An efficient electron microscope can magnify an object by more than one million times its original size. This innovation has thereby allowed scientists to study the precise molecules that constitute human life.

The electron microscope functions by emitting a stream of electrons from a gun-type instrument, which is similar to the apparatus used in an old-fashioned television tube. The electrons pass through an advanced electronic field that is accelerated to millions of volts in certain cases. Before traveling through a vacuum in order to remove oxygen molecules, the electrons are focused into a beam by way of magnetic coils.

Invisible to the naked eye, electron beams can nevertheless be projected onto a florescent screen. When striking the screen, the electrons glow and can even be recorded on film. Cameras also use film to capture images.

In the transmission electron microscope, which is used to study cells or tissues, the beam passes through a thin slice of the specimen that is being studied. On the other hand, in the scanning electron microscope, which is used for tasks such as examining bullets and fibers, the beam is reflected. This reflection creates a picture of the specimen line by line.

43. What is the last step in the process by which the beam emanating from the electron microscope is formed?
    A. The electrons pass through an electronic field.
    B. The electrons are accelerated to millions of volts.
    C. The electrons travel through a vacuum.
    D. Oxygen is removed from the molecules.
    E. The electrons pass through magnetic coils.

44. What is the closest synonym to the word apparatus as it is used in the passage?
    A. machine
    B. electricity
    C. device
    D. tube
    E. bulb

45. Which of the following assumptions has influenced the writer?
    A. The electron microscope has proven to be an extremely important invention for the scientific community.
    B. The invention of the electron microscope would have been impossible without the prior invention of the television.
    C. The electron microscope cannot function without projection onto a florescent screen.
    D. The transmission electron microscope is inferior to the scanning electron microscope.
    E. The electron microscope will soon be an outdated technology.

46. Which statement does not fit the logical flow of the text?
    A. This innovation has thereby allowed scientists to study the precise molecules that constitute human life.
    B. The electrons pass through an advanced electronic field that is accelerated to millions of volts in certain cases.
    C. When striking the screen, the electrons glow and can even be recorded on film.
    D. Cameras also use film to capture images.
    E. This reflection creates a picture of the specimen line by line.

Read the passage below and answer the four questions that follow.

Oliver, having taken down the shutters, was graciously assisted by Noah, who having consoled him with the assurance that "he'd catch it," condescended to help him. Mr. Snowberry came down soon after.

Shortly afterwards, Mrs. Snowberry appeared. Oliver having "caught it," in fulfillment of Noah's prediction, followed the young gentleman down the stairs to breakfast.

"Come near the fire, Noah," said Charlotte. "I have saved a nice little bit of bacon for you from master's breakfast."

"Do you hear?" said Noah.

"Lord, Noah!" said Charlotte.

"Let him alone!" said Noah. "Why everybody lets him alone enough, for the matter of that."

"Oh, you queer soul!" said Charlotte, bursting into a hearty laugh. She was then joined by Noah, after which they both looked scornfully at poor Oliver Twist.

Noah was a charity boy, but not a workhouse orphan. He could trace his genealogy back to his parents, who lived hard by; his mother being a washerwoman, and his father a drunken soldier, discharged with a wooden leg, and a diurnal pension of twopence-halfpenny and an unstable fraction. The shop boys in the neighborhood had long been in the habit of branding Noah, in the public streets, with the ignominious epithets of "leathers," "charity," and the like; and Noah had borne them without reply. But now that fortune had cast his way a nameless orphan, at whom even the meanest could point the finger of scorn, he retorted on him with interest.

Adapted from *Oliver Twist* by Charles Dickens

47. What is the meaning of "he'd catch it" in the first paragraph of the passage?
    A. he'd find it
    B. he'd buy it
    C. he'd be saved
    D. he would be laughed at
    E. he would be punished

48. According to the passage, Oliver could be described as:
    A. gracious
    B. scornful
    C. ignominious
    D. esteemed
    E. ridiculed

49. The passage mainly illustrates:
    A. Charlotte's contempt of orphans.
    B. the wealth of the Snowberry family.
    C. the exploits of Oliver Twist.
    D. Noah's childhood experiences.
    E. the relationship between Noah and Oliver.

50. Who is the "nameless orphan" mentioned in the passage?
    A. charity boys
    B. workhouse orphans
    C. Noah
    D. Oliver
    E. Charlotte

ANSWER KEY AND EXPLANATIONS

Practice Reading Test 3

1. The correct answer is A. You will find the answer in the second item in the index "Exercise: aerobic 243–254." Be careful if you chose answer D. Pages 274–286 are talking about the dangers of exercise in general, not about aerobic exercise in particular.

2. The correct answer is B. You will find the answer in the part of the index on "Exercise: history of 238–242." You can see here that the entries in this section of the index are in chronological order from the 17th century to the 21st century.

3. The correct answer is B. The passage states that gene splicing "has produced results like the super tomato." The phrase "results like" indicates that the super tomato is just one example of this phenomenon.

4. The correct answer is E. Paragraph 1 provides background information on gene splicing, while paragraph 2 gives a brief overview of the ethical and moral debates relating to the topic.

5. The correct answer is C. Please notice the phrase "controversial lately" in the first sentence of paragraph 2, which indicates that there are recent debates about this topic, so the pattern of organization of the passage is "background and recent debates."

6. The correct answer is E. Sentence number 8 contains an opinion when it uses the adjectival phrase "the most important." Answer A is not correct because there would have been historical evidence to show that the friendship between Lewis and Clark existed. Answer B is not correct because this was, in fact, how Lewis was known.

7. The correct answer is D. Sentence 4 comments that Lewis's personality was outgoing and amiable. This is a personal detail which is unrelated to the historical facts in the passage.

8. The correct answer is A. The passage begins with the date of the start of the expedition and talks about the facts that made the expedition possible, so it is giving background information.

9. The correct answer is D. This answer is supported by sentences 3 and 6, which state that Clark was Lewis's trusted friend.

10. The correct answer is E. "Throughout" is the best answer for the first gap because the investigation spanned a certain period of time. "In the end" is the best answer for the second gap because it is to be placed at the beginning of the last sentence of the passage.

11. The correct answer is C. The second to the last sentence of the passage states: "After making inquiries, the FBI discovered that this money originated from the Committee for the Re-election of the President, thereby confirming governmental involvement." Therefore, we can conclude that the Watergate break-in failed because the burglars' money was traced back to a governmental organization, namely the Committee for the Re-election of the President.

12. The correct answer is B. For these types of questions, you need to look for synonyms in the passage. In this passage, "cover-up" is a synonym for "subterfuge," which means concealment.

13. The correct answer is A. This sentence is the best one because the sentences before and after it mention frequencies, so the definition of the word "frequency" is appropriate here.

14. The correct answer is C. The level of vocabulary and technical information in the passage indicate that it is an informative program about brain research. The passage is not technical enough for doctors or surgeons, but it contains too much detail for the general public or for school children.

15. The correct answer is C. The fourth sentence of the passage states: "The work of Berger was confirmed three years later when Edgar Adrian, a Briton, clearly demonstrated that the brain, like the heart, is profuse in its electrical activity." In other words, Adrian confirmed Berger's findings, so Berger paved the way or established a starting point for Adrian.

16. The correct answer is A. The passage implies that governmental funding decisions should primarily be based on the composition of the age-sex groups within its population. This answer is supported by the first sentence in the passage, which states that the age-sex structure is the most significant characteristic of any population.

17. The correct answer is E. The writer gives the examples of elderly citizens and young children in the passage.

18. The correct answer is D. The writer has assumed that society, by way of the government, needs to consider the requirements of all of its members and balance competing needs carefully. This assumption is implied in the last sentence of the passage, which states that "as the composition of a population changes over time, the government may need to re-evaluate its funding priorities."

19. The correct answer is C. Read the graph carefully, do the math, and then check your answers for questions like this one. The library equals 25%, the cafeteria is 18%, the student union is 29%, the sports center is 24%, and other facilities total 4%, so the student union is the highest.

20. The correct answer is B. The passage states: "It took an expedition of Italian scientists, who used a surfeit of technological devices, to disprove Wallerstein's claim." "Plethora," from answer B, means a huge quantity of something. In the passage, a mammoth effort is implied by the phrase "It took an expedition of Italian scientists."

21. The correct answer is E. Paragraph 1 gives the background to the discovery and survey of the peak named Mount Everest, and paragraph 2 talks about disproving Wallerstein's claim, which confirmed the status of Mount Everest as the highest peak in the world. Therefore, "Discovery and survey of Mount Everest" is the best summary of paragraph 1, while "Confirmation of Everest's status" is the best summary of paragraph 2.

22. The correct answer is D. The Italian team confirmed that Everest was, in fact, the tallest mountain in the world. This answer is supported by the last sentence in the passage, which states: "It took an expedition of Italian scientists, who used a surfeit of technological devices, to disprove Wallerstein's claim."

23. The correct answer is D. Many people are not fully aware of certain key information about tornadoes. This lack of awareness is implied by the phrase "myths and misconceptions" in the first sentence of the first paragraph.

24. The correct answer is B. Waterspouts are tornadoes that occur over oceans, rivers, and lakes. See the phrase "bodies of water" in sentence 3 of paragraph 1.

25. The correct answer is B. The phrase "rivers, lakes, and oceans" in answer B ties into the idea of "bodies of water," which is mentioned in the next sentence in the passage.

26. The correct answer is C. Sentence 3 in paragraph 2 gives the recommendation of going to "the underground floor," which means the basement.

27. The correct answer is A. The adjectival phrase "the most influential" in the first sentence indicates that an opinion is being given.

28. The correct answer is B. Answer B is the most general one, and is therefore the best choice. Answer A gives background information, not the central idea. Answer C provides a definition, answer D gives an example, and answer E gives an explanation.

29. The correct answer is A. "Then" is the most suitable for the first gap because the events are being given in chronological order. "As a result" is the best for the last sentence because a conclusion is being stated.

30. The correct answer is C. The last sentence of the passage states that "there has been a surge in artifacts excavated from Africa and Asia." The phrase "has been a surge" indicates that a good deal of development has occurred when compared to earlier centuries.

31. The correct answer is B. The cosmetic use of henna is mentioned in the previous sentence. The phrase "this particular cosmetic" in the sentence after the gap also refers to the cosmetic use of henna.

32. The correct answer is D. Paragraph 1 is devoted exclusively to the topic of eye shadow, while paragraph 2 talks about eye liner, lip and cheek color, nail polish, and hair dye, so the best summaries are "Ancient Egyptian eye shadow" for the first paragraph and "Other ancient cosmetics" for the second paragraph.

33. The correct answer is C. The passage focuses on historical facts, such as Cleopatra's use of cosmetics and the history of the cosmetic use of henna. Thus, the passage is informative in nature.

34. The correct answer is E. The second sentence of the last paragraph mentions that power stations "have been the culprit of the bulk of the acid in rainwater in recent years," meaning that they are the largest contributor to the problem.

35. The correct answer is A. The mention of the chemicals nitrogen and sulfur in paragraph 2 shows that a scientific explanation is being provided. Paragraph 3 talks about "exacerbating the problem," indicating that current problems are being discussed. Accordingly, the writer's approach shifts from scientific explanation to current problems.

36. The correct answer is C. The statement that best supports the writer's main idea is that "the effect of the structures has been to spread the gases more thinly and widely in the atmosphere, thereby exacerbating the problem." This statement links back to the main idea of the passage, which is stated in paragraph 1, sentence 2: "scientific research shows that the acid content of rain has increased dramatically over the past two hundred years, in spite of humanity's recent attempts to control the problem."

37. The correct answer is C. The passage is about music, rather than linguistics, so the sentence stating that "the semantic origins of the word 'monophonic' are interesting to etymologists" is somewhat out of place.

38. The correct answer is B. The academic vocabulary, such as the words "monophonic" and "polyphonic," indicates that the talk is aimed at college students studying music theory. The talk is too formal for an individual music lesson.

39. The correct answer is D. Deciduous trees are those whose leaves change color in the autumn, so chapter 3 will discuss this phenomenon.

40. The correct answer is C. A glossary is the part of a book that provides definitions.

41. The correct answer is E. Sentence 2 mentions what needs to be done to have a moist and fluffy cake. If this advice is not followed, one will get the opposite result, a cake that is dry and dense.

42. The correct answer is B. Be careful when answering questions like this one because the sentences in the passage may not be given in the correct order. Notice the sentence "However, before proceeding with the mixture, ensure that the bowl is of a sufficient size," which indicates that you must check the size of the bowl after preparing the pan and before mixing the ingredients.

43. The correct answer is E. The last step in the process for forming the beam is that the electrons pass through magnetic coils. This answer is found in the last sentence of paragraph 2, which states: "Before traveling through a vacuum in order to remove oxygen molecules, the electrons are focused into a beam by way of magnetic coils." Note that the question focuses on the process of forming the beam in particular, not on the movement of electrons in general, so answer C is incorrect.

44. The correct answer is C. The passage compares the microscope to a television tube in the first sentence of paragraph 2. Both of these items are electronic devices. Machines are larger than devices, so answer A is not the best answer.

45. The correct answer is A. The assumption that has influenced the writer is that the electron microscope has proven to be an extremely important invention for the scientific community. This answer is supported by the second sentence of paragraph 1: "This innovation [i.e., the electron microscope] has thereby allowed scientists to study the precise molecules that constitute human life."

46. The correct answer is D. The statement that "cameras also use film to capture images" does not fit the logical flow of the text. Nothing else about cameras is mentioned in the passage.

47. The correct answer is E. "He'd catch it" means that Oliver was to be punished. This interpretation is supported by paragraph 2, which implies that Mrs. Snowberry is an authoritarian whom the boys fear.

48. The correct answer is E. Noah exclaims about Oliver: "Let him alone!" This exclamation indicates that Oliver is the source of ridicule. The passage also mentions that Oliver is scorned, which is synonymous with being ridiculed.

49. The correct answer is E. The passage mainly illustrates the relationship between Noah and Oliver. This idea is illustrated especially clearly in the last paragraph of the passage, in which we see Noah's view of Oliver.

50. The correct answer is D. The last sentence of the passage states: "But now that fortune had cast his way a nameless orphan, at whom even the meanest could point the finger of scorn, he retorted on him with interest." In this sentence, "his" refers to Noah, so the "nameless orphan" must refer to Oliver.

CBEST Practice Reading Test 4

Read the passage below and answer the three questions that follow.

Equating the whole history of the struggle of humankind to that of the class struggle, the social and political writings of Karl Marx have been the impetus of a great deal of change within society. According to Marxism, the political school of thought based on Marx's doctrines, the working class should strive to defeat capitalism, since capitalistic societies inherently have within them a dynamic that results in the wealthy ruling classes oppressing the masses. The nation state is seen as an instrument of class rule because it supports private capital and suppresses the common person through economic mechanisms, such as the taxation of wages. Because growth of private capital is stimulated by earning profits and extracting surplus value in the production process, wages have to be kept low.

Since capitalism reduces the purchasing power of workers to consume the goods that they produce, Marx emphasized that capitalism inheres in a central contradiction. Under the tenets of Marxism, capitalism is therefore inherently unstable. Marx asserted that productive power ideally should be in the hands of the general public, which would cause class differences to vanish. These idealistic writings have had a huge impact on culture and politics; yet, many believe that Marx's work lacked the practical details needed to bring about the changes to the class structure that he envisaged.

1. The primary purpose of the first paragraph is to:
   A. discuss why the writings of Karl Marx have had such enduring social and political importance.
   B. explain the basic tenets of Marxism, before going on to discuss Marxist views on capitalism and the consequences of private capital.
   C. critique the existing class structure and oppression of the masses.
   D. decry the manner in which taxation and earning profits cause wages to be lowered.
   E. recount the cultural and historical significance of the work of Karl Marx.

2. Which of the following groups of statements best summarizes the main topics addressed in each paragraph?
   A. I. The first paragraph states an assertion; II. The second paragraph refutes that assertion with statistical evidence.
   B. I. The first paragraph explains a long-standing problem; II. The second paragraph provides the potential solution.
   C. I. The first paragraph introduces and expounds upon a theory; II. The second paragraph points out criticisms of the theory.
   D. I. The first paragraph gives the background to the topic in a general way; II. The second paragraph provides specific details about the topic.
   E. I. The first paragraph describes a phenomenon; II. The second paragraph provides further examples of the phenomenon.

3. The writer mentions the "huge impact" that these writings have had on culture and politics in the last sentence in order to:
   A. underscore the fact that class differences have not yet vanished.
   B. highlight the way in which capitalism is often unstable.
   C. reiterate the importance of giving power back to the general populace.
   D. lament the social change that Marx himself predicted.
   E. juxtapose this impact to Marx's failure to include pragmatic instructions in his work.

Read the passage below and answer the three questions that follow.

¹Recent research shows that social media platforms may actually be making us antisocial. ²Survey results indicate that many people would prefer to interact on Facebook or Twitter, rather than see friends and family in person. ³The primary reason cited for this phenomenon was that one does not need to go to the effort to dress up and travel in order to use these social media platforms.

⁴Another independent survey revealed that people often remain glued to their hand-held devices to check their social media when they do go out with friends. ⁵Therefore, social media platforms are definitely having a detrimental impact upon our social skills and interpersonal relationships.

4. In the second paragraph, the author implies that the effect of social media on relationships has largely been:
   A. deleterious
   B. ambivalent
   C. ambiguous
   D. disinterested
   E. objective

5. Which numbered sentence provides an opinion rather than a fact?
   A. Sentence 1
   B. Sentence 2
   C. Sentence 3
   D. Sentence 4
   E. Sentence 5

6. Which of the following, if true, would best support the argument that the author presents in the passage?
   A. Recent research indicates that there has been a rise in antisocial behavior in public places.
   B. Statistics reveal that most children would be reluctant to visit relatives who do not allow the use of smart phones while at the table.
   C. People who feel addicted to their electronic devices may require psychological counseling.
   D. Most people state that they are in control of when and how they use their electronic devices.
   E. Most children would rather be outdoors than watch television.

Read the information below and answer the three questions that follow.

Nassir has a busy work schedule and an active social life. His normal work schedule every week is from Monday to Friday, 8 am to 5 pm. Although it is sometimes necessary, he prefers not to work after 5:00 pm as he likes to visit with friends or take part in his hobbies during the evening. A chart of Nassir's commitments for the month of February is shown below.

Sunday	Monday	Tuesday	Wednesday	Thursday	Friday	Saturday
1	2	3 6:00 pm – personal training at gym	4 7:00 pm – meet Ali in town	5 Staff meeting at work – 1:30 pm	6 8:00 pm – Martial arts practice	7
8	9 Work from 1:00 pm to 9:00 pm this week	10 8:30 am – personal training at gym	11 Staff meeting at work – 3:30 pm	12 9:30 am – personal training at gym	13 10:00 am – meet Marta for brunch	14 8:00 am – Go on hike with Martin and Suki
15 5:00 pm – meet Ali in town	16 8:00 pm – meet Terri for a late supper	17 7:00 pm – Martial arts practice	18 8:30 pm – personal training at gym	19 7:30 pm – event at public library	20 6:00 pm – meet Jacinta for coffee	21 Go to Mom and Dad's this weekend
22 2:00 pm – meet Ali in town	23 Work from 1:00 pm to 5:00 pm this week	24 6:30 pm – meet Tom for supper	25 6:00 pm – meet Jacinta for coffee	26 Staff meeting at work – 2:00 pm	27 8:00 am – Martial arts practice	28 10:00 am – Meet Jim for bike ride

7. Which of the following statements is best supported by the information provided above?
   A. Nassir has staff meetings once a week.
   B. Nassir normally sees his parents once a month.
   C. Nassir finds it difficult to work late in the evening.
   D. Nassir never has to work on Saturday or Sunday.
   E. Nassir writes his work hours on his calendar only when they vary from 8 am to 5 pm.

8. Based on the information provided above, which of the following statements best explains how Nassir plans the time that he spends exercising?
   A. Nassir's gym is open 24 hours a day, so he can go there any time.
   B. Nassir can go to martial arts practice during the evening only.
   C. Nassir does not like to exercise early in the morning.
   D. Nassir likes to exercise with friends when possible.
   E. Nassir never exercises with Ali in town.

9. During the month of February, which of the following activities did Nassir take part in at least four times? You may choose more than one answer.
   A. Meeting friends for supper.
   B. Meeting friends for coffee or brunch.

C. Meeting friends in town.
D. Personal training at the gym.
E. Martial arts training.

Read the passage below and answer the four questions that follow.

How is civil order maintained within any given population? The civil order control function suggests that public order is best maintained through agencies other than the police force or militia. Accordingly, martial law, the establishment of military rule over a civilian population, is only imposed when other methods of civil control have proven ineffective. Either the leader of a country's military system or of the country's own government may lay down the edict for the rule of martial law. In the past, this state of affairs most commonly occurred to quell uprisings during periods of colonial occupation or to thwart a coup, defined as an illegal and usually violent seizure of a government by a select group of individuals.

So, how is the declaration of martial law currently regulated? The constitutions of many countries now make provisions for the introduction of martial law, allowing it only in cases of national emergency or in the case of threats to national security from foreign countries. In democratic nations, severe restrictions are imposed on the implementation of martial law, meaning that a formal declaration of military rule over a nation should be rendered virtually impractical. In much the same way, restrictions have been placed on voting in elections in many countries.

10. Which of the following statements gives the best summary of the main point of the lecture?
    A. In the past, the militia was not used to support civil authorities, although it is used this way at present.
    B. In the past, countries did not have constitutions or other established means to regulate the declaration of martial law.
    C. There are more threats to national security nowadays than there were in the past.
    D. Civil order was more difficult to maintain in the past than it is during the present time.
    E. Certain countries have not had to make declarations of martial law in the past.

11. Which of the following is the best meaning of the word quell as it is used in this passage?
    A. suppress
    B. oppose
    C. obliterate
    D. militarize
    E. bring about

12. What inference about coups can be drawn from the passage?
    A. They usually represent a large proportion of the population.
    B. They no longer occur as countries now have controls in place to prevent them.
    C. They may involve harming government leaders, officials, or citizens.
    D. They have taken place most frequently during periods of colonial occupation.
    E. Controlling coups is virtually impractical nowadays.

13. Which sentence is least relevant to the main idea of the passage?
    A. The civil order control function suggests that public order is best maintained through agencies other than the police force or militia.
    B. Either the leader of a country's military system or of the country's own government may lay down the edict for the rule of martial law.
    C. The constitutions of many countries now make provisions for the introduction of martial law, allowing it only in cases of national emergency or in the case of threats to national security from foreign countries.

D. In democratic nations, severe restrictions are imposed on the implementation of martial law, meaning that a formal declaration of military rule over a nation should be rendered virtually impractical.
E. In much the same way, restrictions have been placed on voting in elections in many countries.

Read the passage below and answer the three questions that follow.

Do mice really prefer cheese to all other foodstuffs? One well-known American exterminator has revealed his secret to catching these pesky rodents: lemon-flavored candy. It appears that the confection has a double advantage. Its sweet smell attracts the mouse much more strongly than does cheese, and its sticky consistency helps to hold the creature captive for the moment it takes for the trap to release.

_____, through logical analogy, we can conclude that it is fallacious to presume that other groups of animals have preferences for certain food groups. _____, we cannot readily conclude that all dogs would choose meat or that all cats would select milk as their favorite foodstuffs.

14. Which of the words or phrases, if inserted in order into the blanks of the passage, would help the reader better understand the progression of the author's argument?
    A. In the same way; Despite this
    B. Therefore; For instance
    C. In this case; On the contrary
    D. On the other hand; Hence
    E. Rather; In effect

15. Which of the following, if true, would fail to support the conclusion that the author makes in the passage?
    A. Mice are attracted more to the texture of the candy than to its smell.
    B. Some animals have a very acute sense of smell.
    C. Many scientific experiments demonstrate that dogs do not prefer the taste and texture of meat to the taste and texture of other food.
    D. Independent observations reveal that mice eat cheese as often as they lemon-flavored candy when both foodstuffs are available to them at the same time.
    E. Quite a few cats are allergic to milk and other dairy products.

16. What is the pattern of organization of the passage?
    A. general to specific
    B. cause and effect
    C. explanation and example
    D. argument and counterargument
    E. theoretical development and new innovations

Read the passage below and answer the three questions that follow.

Ludwig von Beethoven was one of the most influential figures in the development of musical forms during the Classical period. Born in Bonn, Germany, the composer became a professional musician before the age of 12. _____ .

After studying under both Mozart and Haydn, Beethoven became a virtuoso pianist and had many wealthy patrons, who supported him financially. His most popular works are considered to be his fifth and sixth symphonies, and his only opera is entitled *Fidelio*. Listening to any of his creations, it is obvious that that his compositions express the creative energy of the artist himself, rather than being written to suit the demands of his patrons.

17. The main purpose of the passage is to:
    A. suggest that the works of Beethoven, Mozart, and Haydn are very similar.
    B. explore the development of musical composition during the Classical period.
    C. provide background information about Beethoven's life and work.
    D. explain how Beethoven acquired many wealthy patrons.
    E. illustrate the creative energy in Beethoven's symphonies and operas.

18. Which sentence if placed in the line in paragraph 1 would be the most consistent with the writer's purpose and audience?
    A. However, certain people do not believe that Beethoven is worthy of such attention.
    B. His young age masked a variety of emotional and personal problems.
    C. This early started marked the incipience of a very prolific and successful career as a musical artist.
    D. On the other hand, other composers were unable to match this level of success.
    E. This displayed the trust his parents had placed in him.

19. Which of the following assumptions has most influenced the writer?
    A. Beethoven was an incredible, expressive artist.
    B. The work of Mozart and Haydn are often overlooked.
    C. The German composers were more expressive than those from other countries.
    D. Composers would not have been able to write music without their patrons.
    E. Patrons wielded too much control over composers.

Read the passage below and answer the four questions that follow.

Today archeologists are still endeavoring to uncover the secrets of Africa's past. Evidence of the earliest human activity has been found in the south and east of the continent, where climatic conditions helped to preserve the human skeletons and stone tools found there. Genetic science confirms that these are quite likely the oldest remains in the world of modern people, with this classification based on the ability of humans to become adaptable and ready to respond to environmental change.

Even though the artifacts and skeletons of early Africans are most commonly found in a highly fragmented state, these findings are more than sufficient in order to make a number of significant conclusions. Perhaps the most important discovery is that there is great <u>variation</u> among the human remains, indicating a wide array of physical differences among members of the population. While the early population was diverse, it has been well established that the earliest species of hominids spread from Africa to other continents

20. Which of the following best describes the organization of the passage?
    A. A common fallacy is described, and then it is refuted.
    B. An unresolved question is posed, and then it is answered.
    C. A problem is described, and then a solution is discussed.
    D. General information about the research is provided, and then the specific findings of the research are presented.
    E. An archeological method is analyzed, and then a contributing phenomenon is discussed.

21. The author states that "genetic science confirms that these are quite likely the oldest remains in the world of modern people" in paragraph 1 primarily in order to emphasize:
    A. the depth and breadth of Africa's history.
    B. the way that climatic conditions can help to preserve skeletons.
    C. the importance of the stone tools found in African sites.
    D. the significance of archeological discoveries in Africa.
    E. the fact that human beings are adaptable and responsive.

22. As used in paragraph 2, the word variation most likely means:
    A. distinction
    B. discrepancy
    C. possibility for error
    D. changeability
    E. variety

23. From the passage, it can be inferred that some of the archeological discoveries from Africa:
    A. were broken into small pieces or extremely damaged.
    B. would not have been located without modern genetic science.
    C. were not as important as those from other continents.
    D. could not be confirmed by genetic science.
    E. were from inhabitants originally from other continents.

Read the passage below and answer the four questions that follow.

A complex series of interactive patterns govern nearly everything the human body does. We eat to a rhythm, and drink, sleep, and even breathe to separate ones. Research shows that the human body clock is affected by the three main rhythmic cycles of the earth, moon, and sun. These rhythms create a sense of time that is both physiological as well as mental. Humans feel hungry about every four hours, sleep about eight hours in every 24-hour period, and dream in cycles of approximately 90 minutes each.

These natural rhythms, sometimes called circadian rhythms, are partially controlled by the hypothalamus in the brain. Circadian rhythms help to explain the "lark vs. owl" hypothesis. Larks are those who quite rightly prefer to rise early in the morning and go to bed early, while owls are those who feel at their best at night and stay up too late. These cycles explain the phenomenon of jet lag, when the individual's body clock is out of step with the actual clock time in his or her new location in the world. In humans, births and deaths also follow predictable cycles, with most births and deaths occurring between midnight and 6:00 am.

24. In the first paragraph, the author suggests that our mental and physiological sense of time is:
    A. appropriate
    B. exaggerated
    C. oversimplified
    D. overgeneralized
    E. hypochondriacal

25. In the second to the last sentence of the passage, the phrase "these cycles" refers to:
    A. the "lark vs. owl" hypothesis
    B. circadian rhythms
    C. the hypothalamus in the brain
    D. the individual's body clock
    E. cycles of birth and death

26. The author's attitude toward owls in the "lark vs. owl" hypothesis can best be described as one of:
    A. disapproval
    B. skepticism
    C. hostility
    D. support
    E. fanaticism

27. The author would most likely recommend that sufferers of jet lag do which of the following?
    A. Better control their circadian rhythms.
    B. Take medicine to regulate the hypothalamus.
    C. Go to bed earlier than usual.
    D. Stay up later than usual.
    E. Allow their body clocks to adjust to the time difference naturally.

Read the passage below and answer the three questions that follow.

Depicting the events of a single day, James Joyce's epic novel *Ulysses* took more than 20,000 hours, or a total of eight years, to write. Set in Dublin, the novel was initially published in installments as a series before the Parisian publishing house Shakespeare and Company issued a limited edition of 1,000 copies.

The book was risqué for its time, and was classified as obscene material in the United Stated. After the work was cleared of obscenity charges, an unexpurgated version was accepted for publication by Random House in New York. Ironically, it was not available in Dublin until 40 years later.

28. From this passage, it seems safe to conclude which of the following?
    A. Irish publishing companies often engage in dilatory practices when dealing with their authors.
    B. Irish publishers were dissuaded in publishing the novel since it depicted the events of only one day.
    C. Random House did not have a division in Dublin at that time.
    D. Social mores in Dublin were much stricter than those of the United States at that time.
    E. Dublin had a more liberal society than that of Paris.

29. What is the most appropriate title of the passage?
    A. The Life and Work of James Joyce
    B. Problems with James Joyce's *Ulysses*
    C. The Publication of James Joyce's *Ulysses*
    D. A Comparative Analysis of the Writings of James Joyce
    E. How James Joyce Inspired Young Writers

30. Who is most likely to be the audience of this passage?
    A. Young children in an elementary school assembly
    B. Authors attending a conference
    C. Participants in a workshop for writers
    D. College students in a literature lecture
    E. Publishers attending a seminar

Read the passage below and answer the four questions that follow.

¹For every building that is successfully constructed, there are countless others that have never received the chance to leave the drawing board. ²Some of these unbuilt structures were practical and mundane, while others expressed the flights of fancy of the architect. ³Known to us today only through the plans left on paper, many unbuilt buildings were originally designed to commemorate particular people or events.

⁴Such was the case with the monument dubbed the *Beacon of Progress*, which was to be erected in Chicago to display exhibits dedicated to great Americans in history. ⁵In fact, scholar Samantha Mulholland points out that other proposed projects were even more quixotic, like that of *The Floating Spheres*, described as modules held aloft by hot air to house cities of the future.

31. Samantha Mulholland suggests that which of the following explains why some proposed projects were never constructed?
    A. Some projects were never undertaken due to the fact that they did not commemorate any significant event.
    B. The plans for some projects had serious design flaws.
    C. Some projects were too extravagant and impractical ever to be built.
    D. People were not ready to face the future of housing at the time that the construction of *The Floating Spheres* was proposed.
    E. *The Floating Spheres* would have been built had it included an important monument.

32. Which numbered sentence from the passage best introduces the writer's main idea?
    A. Sentence 1
    B. Sentence 2
    C. Sentence 3
    D. Sentence 4
    E. Sentence 5

33. What word best describes the style of writing in this passage?
    A. persuasive
    B. critical
    C. informative
    D. condemning
    E. neutral

34. Between paragraphs 1 and 2, the writer's approach shifts from:
    A. advantages to disadvantages
    B. cause to effect
    C. general to specific
    D. problem to solution
    E. support to criticism

Read the passage below and answer the three questions that follow.

The ancient legal code of Babylonia had severe sanctions for a wide range of crimes. The consequences of crime were best viewed as a way to express personal vengeance. _____, punishments included cutting off the fingers of boys who had hit their fathers or gouging out the eyes of those who had blinded another person. As with most ancient peoples, the Babylonians did not believe in humane treatments for offenders. They also did not have an advanced system of education in those days.

_____ , Sumerian King Ur Nammu, who formulated a set of laws that were surprisingly modern in their approach, did not follow these draconian forms of retribution. Sumerian law stipulated that perpetrators of violent crimes pay monetary damages to their victims, and Ur Nammu's system is the first recorded example of financial awards being imposed in lieu of other forms of punishment.

35. Which of the words or phrases, if inserted in order into the blanks of the passage, would help the reader better understand the progression of the author's argument?
    A. Certainly; In fact
    B. Rather; For instance
    C. In this case; On the contrary
    D. On the other hand; Hence
    E. For example; In contrast

36. The author mentions Sumerian King Ur Nammu primarily in order to:
    A. criticize previous Babylonian rulers.
    B. emphasize the severity of the Babylonian system of justice.
    C. imply that Babylonian sanctions were just for their time.
    D. point out that humane treatments were not really suitable during that epoch.
    E. provide a contrast with the forms of punishment meted out by the Babylonians.

37. Which sentence is least relevant to the main idea of the passage?
    A. The ancient legal code of Babylonia had severe sanctions for a wide range of crimes.
    B. The consequences of crime were best viewed as a way to express personal vengeance.
    C. As with most ancient peoples, the Babylonians did not believe in humane treatments for offenders.
    D. They also did not have an advanced system of education in those days.
    E. Sumerian law stipulated that perpetrators of violent crimes pay monetary damages to their victims, and Ur Nammu's system is the first recorded example of financial awards being imposed in lieu of other forms of punishment.

Read the passage below and answer the three questions that follow.

In his book *Il Milione*, known in English as *The Travels of Marco Polo*, the intrepid explorer describes the marvels he encountered as he journeyed to China. Upon his visit to the emperor Kublai Khan in Cathay, Polo witnessed the magical illusions performed by the court wizards of the supreme ruler. Polo first saw a row of golden cups <u>levitate</u> over the table as Khan drank from each one without spilling a drop. Next, unrestrained lions would appear to lie down in humility in front of the emperor.

However, Khan was venerated for much more than these acts of mere wizardry. The ruler was also regarded for establishing paper currency, a vast postal system, and a well-maintained network of roads. Some academics have disputed the veracity of Polo's written account of the Khan Empire. However, common sense tells us that there would have been little motive for the explorer to have exaggerated his version of events since he was being held captive at the time with no hope of release.

38. It can be inferred from the passage that the primary reason why the court wizards performed magical illusions was to:
    A. venerate the majesty of Kublai Khan.
    B. play a trick on Marco Polo.
    C. provide an interesting story for the book *Il Milione*.
    D. make Kublai Khan and his court appear powerful and mysterious.
    E. help the ruler control wild animals.

39. Which of the following is the best meaning of the word <u>levitate</u> as it is used in this passage?
    A. rise
    B. drag
    C. hover
    D. linger
    E. hang up

40. Which of the following best describes the organization of the passage?
    A. It discusses a problem and then provides a solution.
    B. It recounts a story and then offers an explanation.
    C. It describes a social phenomenon and then illustrates it.
    D. It gives the historical background to a piece of writing and then provides further details about it.
    E. It compares one version of a historical event to a differing account and interpretation of the event.

Read the passage below and answer the three questions that follow.

Two original forms of theater have emerged from Japanese culture: Noh and Kabuki. Noh, the older form, was originally established to meet the demands of the discriminating Japanese aristocracy and remained unchanged for more than six centuries. Noh renders mundane, everyday activities, like drinking tea or arranging flowers, into exquisite artistic performances. With little spoken dialogue, Noh is classified as more ritual than drama.

Kabuki performances are discernably different than those of Noh. Based on puppet theater, Kabuki is designed to meet the tastes of the general populace, rather than those of the aristocracy. According to long-standing theatrical custom, Kabuki performances can be extremely long, lasting up to twelve hours in some cases. Because of its appeal to the general populace, Kabuki theater remains as fascinating and exotic as it has always been.

41. The passage suggests which of the following about followers of Noh?
    A. They lament the fact that Noh clings on to outdated customs of the past.
    B. They believe that Kabuki theater is overtly flamboyant.
    C. They fear that the popularity of Kabuki theater may diminish the appeal of Noh.
    D. They plan to make Noh more up-to-date in order to increase its following.
    E. They want to emphasize that followers of Noh are traditional and discerning.

42. The passage implies that Japanese audiences today would respond to Kabuki theater with:
    A. admiration
    B. impatience
    C. confusion
    D. boredom
    E. detachment

43. Followers of Noh and followers of Kabuki would probably agree with which one of the following statements?
    A. Theatrical productions sometimes last too long.
    B. Japanese theater is unlikely to change in the future.
    C. Theatrical performances should be highly stylized and full of spectacle in order to be effective.
    D. Japanese theater is an important and interesting aspect of Japanese culture.
    E. Japanese theater has been compromised through exposure to other cultures.

Read the passage below and answer the three questions that follow.

Dance notation for choreography can be based on drawings, stick figures, abbreviations, musical notes, or abstract symbols. _____ _____. In the seventeenth century, Pierre Beauchamp devised a notation system for Baroque dance. Known as Beauchamp-Feuillet notation, his system was used to record dances until the end of the eighteenth century. Later, Vladimir Ivanovich Stepanov, a Russian, was responsible for notating choreographic scores for the famous *Sergevev Ballet Collection*.

Two other notation systems, Labanotation and Benesh notation, also known as choreology, are in wide-spread use today. Apple created the first computerized system to display an animated figure on the screen that illustrated dance moves, and many other software systems have been developed to facilitate computerized dance notation.

44. According to the passage, Beauchamp-Feuillet notation differs from Vladimir Ivanovich Stepanov's notation system in that Stepanov's system:
    A. was used for a different genre of dance during a different time period.
    B. was used for works performed in the Russian language.
    C. is now computerized.
    D. was also known as choreology.
    E. did not use animation.

45. Which sentence if placed in the line in paragraph 1 would be the most consistent with the writer's purpose and audience?
    A. However, some choreographers regard the use of stick figures as childish and unprofessional.
    B. Recording the movements of dance through a shortened series of characters or symbols, more than one hundred systems of dance notation have been created over the past few centuries.
    C. As sometimes happens, choreographers cannot read orchestral musical scores.
    D. Nevertheless, other innovations in choreography were more significant than dance notation.
    E. Dance notation is defined as any system that records the individual movements of a dance.

46. The author most likely mentions Apple and other computerized dance notation systems in the last sentence of the passage in order to:
    A. point out the similarities among various computerized systems.
    B. advocate the use of computer software for choreographed performances.
    C. identify the systems that have replaced choreology.
    D. imply that computerized dance notation systems are of better quality than those of the past.
    E. indicate possible trends in dance notation.

Read the passage below and answer the two questions that follow.

Known as the Centennial State, Colorado is divided into sixty-three counties. Colorado joined the union as the 38th state in 1876, shortly after the first substantial discovery of gold in the state near Pikes Peak in 1859. The Rocky Mountains run along a north-south line through the center of the state, and there are several other famous national parks and monuments nearby.

Agriculture in the state involves the production of wheat, hay, corn, sugar beets, and other crops, as well as cattle ranching and raising other livestock. However, the packaging, processing, fabrication, and defense industries form the lion's share of revenues from business and commerce in the state.

47. Which of the following is the best meaning of the words lion's share of as it is used in this passage?
    A. preponderance of
    B. superfluous
    C. excessive
    D. abundant
    E. ever-increasing

48. The primary purpose of the passage is to:
    A. discuss trade and commerce in a particular state.
    B. sum up the historical background and notable features of a particular state.
    C. provide pertinent political details about the acquisition of a particular state.
    D. emphasize the importance of agriculture for trade and commerce in Colorado.
    E. reveal a surprise fact about a particular state.

Read the passage below and answer the two questions that follow.

The Earth's only natural satellite, the Moon lacks its own atmosphere and is only about one-fourth the size of the planet it orbits. The equality of its orbital rate to that of the Earth is the result of gravitational locking, also known as synchronous rotation. Thus, the same hemisphere of the Moon always faces the earth. Scientific dating of samples from the Moon's crust reveals that the materials range in age from three to four billion years old.

Lunar evolution models suggest that the development of the Moon occurred in five principle stages: (1) increase in mass followed by large-scale melting; (2) separation of the crust with concurrent bombardment by meteors; (3) melting at greater depth; (4) lessening of meteoric bombardment with further melting at depth and the formation of basaltic plains; and (5) the cessation of volcanic activity followed by gradual internal cooling.

49. Which of the following situations best further illustrates the concept of synchronous rotation?
    A. The Moon goes through four phases every twenty-eight days.
    B. A star appears to shine at the same intensity, regardless of its position in the sky.
    C. Two objects fall to the ground at the same speed and land at the same time.
    D. An undiscovered planet has two equal hemispheres.
    E. A telecommunications satellite is always in the same position above a certain city on Earth.

50. The passage suggests that which one of the following probably occurred after the completion of the process of lunar evolution?
    A. Ice continued to melt on the surface of the Moon.
    B. The likelihood of the collision of the Moon with a meteor was substantially reduced.
    C. Melting at depth still occurred.
    D. There were further eruptions of magma or lava.
    E The temperature of the internal core of the Moon was lower than it was previously.

ANSWER KEY AND EXPLANATIONS

Practice Reading Test 4

1. The correct answer is B. The primary purpose of the first paragraph is to explain the basic tenets of Marxism, before going on to discuss Marxist views on capitalism and the consequences of private capital. We know this because the second sentence begins with the phrase "according to Marxism." Answer A is too general, and answers C and D are too specific. Answer E is incorrect because the passage talks about Marxism from a political, rather than a historical perspective.

2. The correct answer is D. The first paragraph gives the background to the topic in a general way, and the second paragraph provides specific details about the topic. You may be tempted to choose answer C, but the criticism is only one aspect of the information provided in paragraph 2.

3. The correct answer is E. The writer mentions the "huge impact" that these writings have had on culture and politics in the last sentence in order to juxtapose this impact to Marx's failure to include pragmatic instructions in his work. We know that the author is making a juxtaposition or comparison because the sentence begins with the word "yet ."

4. The correct answer is E. In the second paragraph, the author suggests that the effect of social media on relationships has largely been deleterious, which means harmful. The author asserts that "social media platforms may be having a detrimental impact upon our social skills and interpersonal relationships."

5. The correct answer is E. Sentence 5 states: "Therefore, social media platforms are definitely having a detrimental impact upon our social skills and interpersonal relationships." The word "definitely" indicates that an opinion is being given. We cannot give a definite conclusion because only two studies have been cited in the article. There are certainly other studies that extol the benefits of social media, but the author has decided not to mention these positive aspects of the topic.

6. The correct answer is B. Statistics revealing that most children would be reluctant to visit relatives who do not allow the use of smart phones while at the table would best support the argument presented in the passage. These statistics would support the claim in the passage that "survey results indicate that many people would prefer to interact on Facebook or Twitter, rather than see friends and family in person."

7. The correct answer is E. From the information provided in the chart, we can conclude that Nassir writes his work hours on his calendar only when they vary from 8 am to 5 pm. The facts at the top of the chart tell us that "his normal work schedule every week is from Monday to Friday, 8 am to 5 pm." He has made notes on his calendar for the week beginning on the 9th and the week beginning on the 23rd, so we can surmise that his normal work schedule has been changed for those two weeks.

8. The correct answer is D. From the information provided, we can conclude that Nassir likes to exercise with friends when possible. On the 14th, he is going on a hike with Martin and Suki, and on the 28th he is meeting Jim for a bike ride.

9. The correct answer is D. During the month of February, Nasir will take part in personal training at the gym four times, on the 3rd, 10th, 12th, and 18th of the month.

10. The correct answer is B. The main point of the lecture is that in the past, countries did not have constitutions or other established means to regulate the declaration of martial law. The second paragraph explains that "the constitutions of many countries now make provisions for the introduction

of martial law." The use of the word "now" suggests that these provisions were not in place in the past.

11. The correct answer is A. "Quell" means to suppress or conquer.

12. The correct answer is C. The last sentence of the first paragraph states that a coup is "defined as an illegal and usually violent seizure of a government by a select group of individuals." The use of the word "violent" suggests that other people may be harmed or even killed

13. The correct answer is E. The following sentence is least relevant: "In much the same way, restrictions have been placed on voting in elections in many countries." The passage focuses on civil order and martial law. The statement about voting is therefore off-topic.

14. The correct answer is B. The first sentence of paragraph 2 makes a conclusion and the second sentence of paragraph 2 gives an example. So, the best choices are "therefore" and "for instance."

15. The correct answer is D. The passage states: "Through logical analogy, we can therefore conclude that it is fallacious to presume that other groups of animals have preferences for certain food groups." This conclusion would not be supported if it were true that mice eat cheese as often as they lemon-flavored candy when both foodstuffs are available to them at the same time. If the mice eat both food groups equally, they would not have a preference, and the logical analogy relies upon the existence of this preference.

16. The correct answer is C. The first paragraph explains why the smell and texture of the candy attract mice. The second paragraph gives the case of dogs' food preferences as a further example.

17. The correct answer is C. The primary purpose of the passage is to provide background information about Beethoven's life and work. The passage begins by providing information about the composer's musical training, before going on to talk about his professional life and compositions.

18. The correct answer is C. The words "prolific and successful career" in the new sentence support the ideas of being a professional at a young and of becoming a virtuoso pianist.

19. The correct answer is A. We know that the author has a high regard for the composer because the passage states: "Listening to any of his creations, it is obvious that that his compositions express the creative energy of the artist."

20. The correct answer is D. General information about the research is provided, and then the specific findings of the research are presented. The first paragraph describes the background to archeological research in Africa, and the second paragraph gives specific details about the remains and artifacts that were discovered there.

21. The correct answer is D. The author states that "genetic science confirms that these are quite likely the oldest remains in the world of modern people" in paragraph 1 primarily in order to emphasize the significance of archeological discoveries in Africa. We know this because the paragraph goes on to explain that "these findings are more than sufficient in order to make a number of significant conclusions."

22. The correct answer is E. As used in paragraph 2, the word "variation" most likely means variety. "Variation" is synonymous with the words "wide array" and "diverse," which are used later in paragraph 2.

23. The correct answer is A. From the passage, it can be inferred that some of the archeological discoveries from Africa were broken into small pieces or extremely damaged. The first sentence of paragraph 2 of the passage tells us that "the artifacts and skeletons of early Africans are most commonly found in a highly fragmented state." "Fragmented" means broken into pieces.

24. The correct answer is A. In the first paragraph, the author suggests that our mental and physiological sense of time is appropriate. The author explains that all human beings have this sense of time. The author does not criticize this behavior, but rather, provides factual information about the topic. From the tone of the passage, we can therefore surmise that that author views this behavior as appropriate.

25. The correct answer is B. In the second to the last sentence of the passage, the phrase "these cycles" refers to circadian rhythms. The two previous sentences state: "Circadian rhythms help to explain the "lark vs. owl" hypothesis. Larks are those who quite rightly prefer to rise early in the morning and go to bed early, while owls are those who feel at their best at night and stay up too late." Larks and owls are given as an example, so the phrase "these rhythms" refers back to the subject of circadian rhythms in the previous sentence.

26. The correct answer is A. The author's attitude toward owls in the "lark vs. owl" hypothesis can best be described as one of disapproval. The author says that larks "quite rightly prefer to rise early in the morning," but owls "stay up too late." So, the author disapproves of the owl's behavior.

27. The correct answer is E. The author would most likely recommend that sufferers of jet lag allow their body clocks to adjust to the time difference naturally. The author begins the second paragraph by explaining that "these natural rhythms, sometimes called circadian rhythms, are partially controlled by the hypothalamus in the brain." Since the author refers to the rhythms as a natural phenomenon, he or she would most likely suggest that the time difference be overcome naturally.

28. The correct answer is D. We can make this conclusion because the author suggests that social mores in Dublin were much stricter than those of the United States at the time that *Ulysses* was published in New York. The passage tells us that "the book was risqué for its time" and was originally classified as "obscene material." In this context, the word "mores" means moral views, and the word "risqué" means indecent.

29. The correct answer is C. Paragraph 1 uses the words "published" and "publishing," while paragraph 2 uses the word "publication." So, the best title is "The Publication of James Joyce's *Ulysses*."

30. The correct answer to D. This would likely be introductory material in a literary lecture because the passage gives some basic background to the book. The passage does offer any advice that would be relevant to writers, authors, or publishers.

31. The correct answer is C. Samantha Mulholland suggests some proposed projects were never constructed because they were too extravagant and impractical ever to be built. The passage states: "Scholar Samantha Mulholland points out other proposed projects were far more quixotic." The word "quixotic" means extravagant and impractical.

32. The correct answer is A. The rest of the passage talks about buildings that "never received the chance to leave the drawing board."

33. The correct answer is C. The passage merely provides information. It does not try to persuade, nor does it give any opinions.

34. The correct answer is C. Paragraph 1 gives the general background to the topic, while paragraph 2 gives two specific examples of unbuilt structures.

35. The correct answer is E. The third sentence of paragraph 1 gives an example of a punishment. The first sentence of paragraph 2 contrasts these punishments to a different set of laws. So, the best answers are "for example" and "in contrast."

36. The correct answer is E. The author mentions Sumerian King Ur Nammu primarily in order to provide a contrast with the usual forms of punishment meted out by the Babylonians. The passage states that "the Babylonians did not believe in humane treatments for offenders." However, King Ur Nammu "did not follow these draconian forms of retribution."

37. The correct answer is D. The passage is describing laws, so the comment about education is not relevant.

38. The correct answer is D. It can be inferred from the passage that the primary reason why the court wizards performed magical illusions was to make Kublai Khan and his court appear powerful and mysterious. The first paragraph uses the words "amazement" and "astonishing" to express the mysteriousness of the court.

39. The correct answer is C. The author most probably uses the word "levitate" in paragraph 1 to mean hover. The words "levitate" and "hover" both mean to be suspended in midair.

40. The correct answer is D. The passage gives the historical background to a piece of writing and then provides further details about it. Paragraph 1 describes the book *Il Milione*, and paragraph 2 provides some additional information about Polo's written account of events.

41. The correct answer is E. The passage suggests that the followers of Noh are traditional, discerning, and serious. Paragraph 1 states that Noh is for the "discriminating Japanese aristocracy." The word "aristocracy" indicates that the dance is traditional in nature, while the word "discriminating" means "discerning."

42. The correct answer is A. The second paragraph implies that Japanese audiences today would respond to Kabuki theater with admiration. The last sentence of the second paragraph states: "Because of its appeal to the general populace, Kabuki theater remains as fascinating and exotic as it has always been." We can surmise that people probably admire something that fascinates them.

43. The correct answer is D. Followers of Noh and followers of Kabuki would probably agree that Japanese theater is an important and interesting aspect of Japanese culture. The first sentence of the passage explains that these forms of theater "have emerged from Japanese culture." Since an article has been devoted to this topic, we can assume that followers consider the topic to be an important and interesting aspect of the Japanese culture.

44. The correct answer is A. According to the passage, Beauchamp-Feuillet notation differs from Vladimir Ivanovich Stepanov's notation system in that Stepanov's system was used for a different genre of dance during a different time period. Stepanov's system was used for ballet after the eighteenth century. The system that Pierre Beauchamp devised was used for Baroque dance until the end of the eighteenth century.

45. The correct answer is B. The phrase "past few centuries" in the new sentence provides a link to the historical background provided in the next paragraph. You may be tempted to choose answer choice E, which offers a definition. However, answer choice B is better because it has both a definition and the chronological connection.

46. The correct answer is E. The author most likely mentions Apple and other computerized dance notation systems in the last sentence of the passage in order to indicate possible trends in dance notation. We know this because the sentence focuses on new developments in dance notation.

47. The correct answer is A. The phrase "preponderance of" could be substituted for "lion's share of" in the second paragraph with the least change in meaning. Both phrases refer to the majority of something.

48. The correct answer is B. The primary purpose of the passage is to sum up the historical background and notable features of a particular state. The other answer choices provide specific details from the passage, rather than the primary purpose.

49. The correct answer is E. The concept of synchronous rotation is illustrated in the situation in which a telecommunications satellite is always in the same position above a certain city on Earth. This is similar to the way in which the same hemisphere of the Moon always faces the earth.

50. The correct answer is E. Point 5 in paragraph 2 states that the last step in lunar evolution was "the cessation of volcanic activity followed by gradual internal cooling." So, we can conclude that after lunar evolution, the temperature of the internal core of the Moon was lower than it was previously.

CBEST ESSAY WRITING

PART 1 – ABOUT THE CBEST ESSAYS

CBEST Essay Format & Question Types

CBEST essay writing

Your CBEST test will include a written essay.

The purpose of the essay is to assess your ability to express and develop your thoughts in writing.

The essay assesses this skill because clear writing is essential for a successful teaching career.

Test administration

You may be asked to write your essay on paper or on a computer.

If you are asked to write your essay on a computer, you can ask for scratch paper to take notes and plan your essay.

Study aids, such as dictionaries or grammar books, are not permitted.

Time limit

Normally, you will be given 30 minutes to plan, write, and edit each of your two CBEST essays.

Word count

Although there is no official word limit, your essays should normally be between 300 to 450 words each.

The CBEST Expository Essay Task

Your first CBEST essay question is an expository task.

For the expository task, you will have to analyze a statement, and then give reasons for the opinions you have expressed.

In other words, you will need to take a stand on the issue presented and support your viewpoint with reasons and examples.

The expository essay is designed to allow you to demonstrate your analytical skills.

Note that you will not be expected to demonstrate specialist knowledge in any particular academic subject area for your expository essay response.

Here is a sample expository essay question:

> Most Americans have access to computers and cell phones on a daily basis, making email and text messaging extremely popular. While some people argue that email and texting are now the most convenient forms of personal communication, others believe that electronic communication technology is often used inappropriately. Write an essay for an audience of educated adults in which you take a position on this topic. Be sure to provide reasons and examples to support your viewpoint.

We will see sample responses to this essay topic in the second part of the essay writing study guide.

The CBEST Personal Experience Essay Task

The second CBEST essay task is a personal experience piece.

You will be required to write about a personal experience you have faced.

The personal experience essay topic is designed to elicit expressive writing about an experience that you remember well.

In other words, the aim of this task is expressive writing, so you can make your tone much more personal than in essay 1.

For this reason, most students find essay task 2 much easier to write than essay task 1.

If permitted to do so, you may therefore wish to write task 2 before attempting task 1 on the day of your actual CBEST test.

Note that you will not be asked to write letters or other forms of correspondence for the CBEST personal experience essay.

You will also not be asked to write about a hypothetical or imaginary situation for this essay task.

We will see sample personal experience essay topics and model responses in the third part of the study guide.

Essay Scoring – How Your CBEST Essays Are Marked

The six following characteristics of your CBEST essays will be assessed:

1. <u>Clear central idea</u> – This means that your essay should answer the question that has been posed. You will need to express your main idea in a clear way in the introduction of the essay. The scorer reading your essay assesses this aspect of your essay by searching for a thesis statement in the first paragraph of your essay.

2. <u>Well-supported</u> – Your essays should demonstrate unity and coherence among the examples that you use to support your argument.

   For the expository essay, you need to be sure that you take a stand on one side of the issue or the other. Your score will not be affected by the position you take.

   For the personal experience piece, it is extremely important to elaborate on the main idea of your essay and maintain your point of view throughout your writing.

   Your essays should include examples and explanations that illustrate and support your viewpoints.

   The scorer reading your essays searches for linking words and phrases that signal that examples or reasons are being provided in the essay. These linking words and phrases include the following: "such as," "for example," "for this reason," and "because of."

3. <u>Logical organization</u> – Your essay should be divided into paragraphs, which have been set out in an organized manner. Each body paragraph should contain a point that supports your main idea. You should also include a conclusion that sums up the essay.

The scorer reading your essay looks for logical paragraph divisions, as well as for linking words and phrases which indicate that a new paragraph is beginning.

4. <u>Writing conventions</u> – Your essay should be grammatically accurate and punctuated correctly. Your spelling should also be correct.

5. <u>Syntactic complexity</u> – You should write long and developed sentences that demonstrate a variety of sentence patterns. You should avoid repeatedly beginning your sentences in the same way, such as "I think that."

   The scorer reading your essay will look for a variety of sentence patterns.

6. <u>Appropriate tone and style</u> – Your essay needs to address the concerns of your target audience.

   You need to be sure that you have used the correct word choice and style in order to achieve this purpose.

   Generally speaking, the tone of your expository essay should be formal, while the tone of your personal experience essay should be expressive.

How to Avoid Common Essay Errors and Raise Your Score

In the previous section, we talked about the characteristics of a well-written essay.

However, you may also wonder which aspects of an essay would be scored poorly by the person who is evaluating your written work.

These errors most commonly cause students to receive a low score on the CBEST essay:

1. The essay fails to express a clear point of view or provides a viewpoint that cannot be logically supported.
    Tip: You can avoid this error by giving a clear thesis statement in the first paragraph of your essay.

2. The essay is written in a tone and style that is not suitable for the audience.
    Tip: Using the correct tone and style involves avoiding slang expressions in your writing. Examples of slang language are words like "awesome" or "guy."

3. The reasons or examples provided in the essay are flawed because they do not support the student's main point.
    Tip: Be sure that your reasons and examples are closely related to your main idea and to the essay topic. For instance, if you are asked whether art programs should be supported in schools, and then go on to talk about physical education programs because you believe they are similar to art programs, your reasoning would be flawed.

4. The essay is disorganized and therefore difficult to read and score.
    Tip: You can avoid this error by brainstorming your ideas and planning your essay before you begin writing.

5. The essay contains errors in sentence construction or contains only simple or repetitive sentence structures.
    Tip: Try to avoid writing every sentence of your essay in the subject-verb-object sentence pattern. In order to avoid this shortcoming, you can begin sentences with words and phrases like "although" or "because of this."

6. The essay does not demonstrate a complex thought process.
    Tip: Be sure that you give reasons and examples to express and support your position.

7. The essay contains errors in spelling, grammar, and punctuation.
    Tip: If you have weaknesses in these areas, you should pay special attention to Parts 4 and 5 of this study guide.

PART 2 – WRITING THE CBEST EXPOSITORY ESSAY

CBEST Expository Essay Structure

Most teachers agree that the best CBEST expository essays follow a four to five paragraph format. This format will help to insure that your essay is well-organized.

This format also helps you write longer and more developed essays that will contain 300 to 450 words.

The five paragraph essay is organized as follows:

Paragraph 1 – This paragraph is the introduction to your essay. It should include a thesis statement that clearly indicates your main idea. It can also give the reader an overview of your supporting points.

Paragraph 2 – The second paragraph is where you elaborate on your first supporting point. It is normally recommended that you state your strongest and most persuasive point in this paragraph.

Paragraph 3 – You should elaborate on your main idea in the third paragraph by providing a second supporting point.

Paragraph 4 – You should mention your third supporting point in the fourth paragraph. This can be the supporting point that you feel to be the weakest.

Paragraph 5 – In the fifth and final paragraph of the essay, you should make your conclusion. The conclusion can sum up your position or leave the reader with an interesting anecdote or example to consider.

Creating Effective Thesis Statements

What is a thesis statement?

A thesis statement is a sentence that asserts the main idea of your essay. For CBEST expository essays, it is recommended to place the thesis statement in the first sentence of the first paragraph of your essay.

This will make your position on the topic clear and will also create a strong impact as the scorer begins to read your essay.

Most expository essays on the CBEST will be on debatable or contentious topics.

You will not need to write about both sides of the argument for the given topic.

You only need to state which side of the argument you support and give reasons for your viewpoint.

Write it early

It is important to draft your thesis statement early in the writing process so that your writing has focus.

However, be prepared to go back and edit your thesis statement after you have finished the main body of your essay.

Keep it focused

Remember that the best thesis statements are those that serve to narrow the focus of the essay and control the flow ideas within it.

As such, a thesis statement should not be too general or vague.

For the CBEST essay, it is recommended to begin your sentence with an identifying phrase, such as "I think that" or "I agree that."

Bearing these tips in mind, you should now complete the following thesis statement exercise.

Thesis Statement – Exercise

Now look at our essay topic below and write a focused thesis statement.

Most Americans have access to computers and cell phones on a daily basis, making email and text messaging extremely popular. While some people argue that email and texting are now the most convenient forms of personal communication, others believe that electronic communication technology is often used inappropriately. Write an essay for an audience of educated adults in which you take a position on this topic. Be sure to provide reasons and examples to support your viewpoint.

Your thesis statement:

Thesis Statement – Answer to Exercise

Suggested answer:

I agree with the assertion that electronic communication technologies such as email and social media platforms are sometimes used inappropriately.

Analysis:

- The thesis statement is clear since it indicates the student's point of view on the topic.
- The student has also identified the sentence as a thesis statement by beginning with the phrase "I agree with the assertion that."
- The above sentence is also an excellent example of a thesis statement since the student focuses his response. The response focuses on email and social media platforms, stating that these media forms are sometimes used inappropriately.

Writing the Introduction

<u>What is the purpose of the introduction?</u>

The purpose of your introduction is to give a brief statement of your point of view and to provide an overview of your supporting points.

<u>What can I include in my introduction?</u>

You can include a vivid example, an interesting fact, a paradoxical statement, or supporting points in your introduction.

<u>When should I write the introduction?</u>

Although it is advisable to write your thesis statement before beginning your main body, you can often go back and write the remainder of the introduction after you have finished the body paragraphs and conclusion.

That is because sometimes it is easier to introduce your essay after you have already written it and developed your points.

<u>What is the structure of the introduction?</u>

The "Assertion + Reason" Structure – A good structure for the introduction is to think of it in terms of an assertion plus a reason or explanation. This structure is better than just giving your assertion or opinion on its own because your explanation indicates the direction that your writing is going to take.

In addition, the "assertion + reason" structure will result in an introduction that contains more words and which is usually richer grammatically and structurally.

The scorer will assess these grammatical and structural aspects of your thesis statement.

Remember that the introduction announces your main idea and supporting points, while your main body develops them.

Writing the Introduction – Exercise

Look at our previous essay topic again and write an introduction of 50 to 75 words. Remember to include your thesis statement at the beginning of your introduction. A sample answer is provided on the following page.

> Most Americans have access to computers and cell phones on a daily basis, making email and text messaging extremely popular. While some people argue that email and texting are now the most convenient forms of personal communication, others believe that electronic communication technology is often used inappropriately. Write an essay for an audience of educated adults in which you take a position on this topic. Be sure to provide reasons and examples to support your viewpoint.

Your introduction:

Writing the Introduction – Answer

Suggested answer:

I agree with the assertion that electronic communication technologies such as email and social media platforms are sometimes used inappropriately. Modern forms of communication such as electronic mail and SMS messaging can cause problems with personal relationships because of three main shortcomings with these media: their impersonal nature, their inability to capture tone and sarcasm, and their easy accessibility at times of anger.

Analysis:

The second sentence of the introduction follows the "assertion + reason" structure.

The assertion is that "Modern forms of communication such as electronic mail and SMS messaging can cause problems with personal relationships."

The reasons are "their impersonal nature, their inability to capture tone and sarcasm, and their easy accessibility at times of anger."

The expository essay will be focused because it will have three main body paragraphs, which will discuss each of the reasons given in the introduction.

Organizing the Main Body

"Expository" means to explain. Therefore, each paragraph of your expository essay should explain and expand on your position on the topic.

Each paragraph of your main body should consist of the following elements:

1. A topic sentence which concisely states the supporting point that you are going to discuss in the paragraph.
2. Well-written and complex sentences that elaborate on your supporting points through reasons and examples.
3. The use of subordination and linking words in order to create a variety of different types of sentence construction. If you are unsure about how to subordinate sentences, you may now wish to turn your attention to Part 5 of the study guide before proceeding.

You may wish to write the body of the paragraph before writing your topic sentence for it.

That is because sometimes it is easier to sum up the main point of the paragraph after you have written it.

For this reason, we will next look at the elaboration of supporting points and writing the main body sentences, before turning our attention to topic sentences.

Elaboration in the Body Paragraphs

<u>How long should each body paragraph be?</u>

For a five paragraph essay, each body paragraph can be approximately 60 to 100 words.

<u>What is an elaborating idea?</u>

Elaborating ideas include both explanations and examples. Providing clear examples to support your points is extremely important.

Each of your main body paragraphs should contain an example that supports your line of argument. You should elaborate on and explain your example in order to make your essay easy to read and follow.

<u>How do elaborating ideas help to raise my CBEST score?</u>

Elaboration lengthens your essay and gives you more opportunities to demonstrate higher-level grammar and complex sentence construction.

<u>How many elaborating ideas should I have in each paragraph?</u>

You should provide two or three elaborating ideas for each body paragraph.

<u>How do I link my elaborating ideas to one another?</u>

You should seamlessly link your elaborating points together to make a coherent paragraph.

This is the function of linking words and subordination, which we will see in Part 5 of the study guide.

<u>How do I come up with elaborating ideas for each supporting point?</u>

Perhaps the best way to elaborate on your supporting points is to take each of the supporting points that you are going to talk about in your main body paragraphs, place them as headings on a piece of scratch paper, and make a list of examples and explanations under each heading.

We will have a look at how to do this in the following exercise.

Elaboration of Supporting Points – Exercise

Let's turn our attention to our sample essay on email and text communication.

Here is the introduction again for ease of reference:

> I agree with the assertion that electronic communication technologies such as email and social media platforms are sometimes used inappropriately. Modern forms of communication such as electronic mail and SMS messaging can cause problems with personal relationships because of three main shortcomings with these media: their impersonal nature, their inability to capture tone and sarcasm, and their easy accessibility at times of anger.

This will be a five paragraph essay, so in your first body paragraph you need to elaborate on the impersonal nature of electronic communication.

Your second body paragraph will elaborate on how emails and texts cannot convey tone and sarcasm.

The third paragraph will talk about the danger of having an accessible messaging service during times of high emotion.

Exercise – Now try to make a list of the ideas you are going to use as elaboration for each of your main body paragraphs. Sample responses are on the following page.

Elaboration – Body Paragraph 1:

Elaboration – Body Paragraph 2:

Elaboration – Body Paragraph 3:

Elaboration of Supporting Points – Answer to Exercise

Elaboration – Body Paragraph 1:

Elaborate on the impersonal nature of electronic communication

- Email is practical, but not always appropriate. Example: informing someone about a death
- No human contact – can be seen as cold or shallow – not like talking on phone or in person

Elaboration – Body Paragraph 2:

Emails and texts cannot convey tone and sarcasm

- It is possible for sarcastic comments to be taken literally
- Message clear to sender, but tone of emotion is conveyed by voice
- Without tone, may come across as demanding, indifferent, etc.

Elaboration – Body Paragraph 3:

The danger of having an accessible messaging service during times of high emotion

- Examples: breaking up with someone by text; firing someone by email
- Easy to send a message quickly when angry – can hurt relationships – waiting and thinking requires self-control & discipline

Writing the Main Body Paragraphs – Exercise

We had a look at brainstorming ideas for your main body paragraphs in the previous section of this study guide. We will now focus on the ideas we developed in that section and write the paragraphs of the main body of our essay. Within each exercise, we reproduce our list of elaborating points for the body paragraphs for ease of reference. If you have difficulties with grammar, this would be a good point to have a look at the CBEST English Grammar Review section in Part 4 of this study guide.

Instructions: Now write the sentences for main body paragraphs for your expository essay, excluding the topic sentences. Some words from the basic sentence structure of the sample response are provided in order to guide you. Refer to the lists above each exercise to help you. Write each new sentence in the order of the points provided.

Main Body Paragraph 1:

Elaborate on the impersonal nature of electronic communication

- Email is practical, but not always appropriate. Example: informing someone about a death
- No human contact – can be seen as cold or shallow – not like talking on phone or in person

Sentence 1: Although email may be practical for . . . , electronic messaging would be remarkably inappropriate for . . .

Sentence 2: There is no direct human contact in . . . , and during times of loss or tragedy, human warmth . . .

Main Body Paragraph 2:

Emails and texts cannot convey tone and sarcasm

- It is possible for sarcastic comments to be taken literally
- Message clear to sender, but tone of emotion is conveyed by voice
- Without tone, may come across as demanding, indifferent, etc.

Sentence 1: For instance, it might be possible . . . of a sarcastic email message to . . .

Sentence 2: The tone of . . . may seem abundantly clear to . . . , but sarcastic or ironically humorous utterances can only . . .

Sentence 3: Without . . . , certain phrases in an email may . . .

Main Body Paragraph 3:

The danger of having an accessible messaging service during times of high emotion

- Examples: breaking up with someone by text; firing someone by email
- Easy to send a message quickly when angry – can hurt relationships – waiting and thinking requires self-control & discipline

Sentence 1: In this day and age, we have heard stories not only of . . . but also of employers who . . .

Sentence 2: Unless the writer of the message has . . . before . . . , he or she might send a regrettable message that can . . .

Suggested Answers – Main Body Paragraphs

Sample Body Paragraph 1 (excluding topic sentence):

Although email may be practical for conveying straightforward information or facts, electronic messaging would be remarkably inappropriate for events like announcing a death. There is no direct human contact in emails and texts, and during times of loss or tragedy, human warmth and depth of emotion can only truly be conveyed through a phone call, or better still, by talking face to face.

Sample Body Paragraph 2 (excluding topic sentence):

For instance, it might be possible for the recipient of a sarcastic email message to take its contents literally. The tone of the message may seem abundantly clear to the person who sent it, but sarcastic or ironically humorous utterances can only really be communicated in speech through the tone and inflection of the voice. Without the aid of tone and inflection, certain phrases in an email may come across as demanding, indifferent, or rude.

Sample Body Paragraph 3 (excluding topic sentence):

In this day and age, we have heard stories not only of personal break ups that have been conducted by text, but also of employers who fire their staff by email message. Unless the writer of the message has the discipline and self-control to give him or herself a period of reasoned contemplation before sending the communication, he or she might send a regrettable message that can cause irretrievable damage to a relationship.

Writing Clear and Concise Topic Sentences

Why are topic sentences important?

As the scorer reads each new paragraph of your CBEST essays, he or she will look for new ideas by searching for words and phrases that you have not used previously in your writing.

Each topic sentence can therefore be a paraphrase, *but should not repeat word for word*, the supporting points in your introduction.

What is the purpose of a topic sentence?

You can think of the topic sentence as a summary of the content of a main body paragraph. The topic sentence serves two purposes.

First of all, it gives an overview of the content of the paragraph because it announces the topic that you are going to discuss.

Secondly, the topic sentence links back to the introduction since it is an elaboration of one of the supporting points that you have already cited at the beginning of the essay.

In this way, clear and concise topic sentences give your essay cohesion and coherence.

Is a topic sentence general or specific in its focus?

While the topic sentence is more specific than the introduction to the essay, the topic sentence should be more general than the elaboration that you are going to make in the paragraph.

In other words, each main body paragraph should move from the more general supporting point that you mention in your topic sentence to the specific points that you raise in your elaboration.

How do I avoid repeating myself?

Remember that although your topic sentences point back to the introduction, you need to avoid using the exact same wording in your topic sentences as in your introduction.

For instance, if you refer to the "impersonal nature of electronic communication" in your introduction, your topic sentence should word this idea differently.

In this case, the phrase "impersonal nature of electronic communication" could be paraphrased by stating: "There is no direct human contact in email."

Where should a topic sentence be placed within the paragraph?

The most common position for the topic sentence is the first sentence of the paragraph.

In longer essays, it is possible to put a topic sentence as the second sentence of a paragraph if the paragraph's first sentence is transitional.

You can also delay the topic sentence until the end of the paragraph for emphasis, although for the sake of clarity, this is not recommended.

For the CBEST essay, you should plan to write three or four main body paragraphs, each of which have a topic sentence as their first sentence.

Topic Sentences – Exercise

Let's turn our attention to writing topic sentences for the main body paragraphs of our essay on email and text communication.

The topic sentence for your first body paragraph will mention the impersonal nature of electronic communication.

Your topic sentence for the second body paragraph will mention how the tone of emails and texts can be misunderstood.

The topic sentence of your third paragraph will talk about the danger of having a quick messaging service at hand when you are angry.

Exercise – Now try to write topic sentences for each of your main body paragraphs. Remember that the topic sentence needs to be specific for each supporting point, but general enough to introduce the paragraph as a whole. You may wish to refer back to what you have written thus far on the essay topic. Sample responses are provided on the next page.

Topic Sentence 1:

Topic Sentence 2:

Topic Sentence 3:

Topic Sentences – Answer to Exercise

Here are possible topic sentences for the three main body paragraphs.

Topic Sentence 1:
Depending upon the context, the recipient of an email or text message may consider this mode of communication to be insensitive or uncaring.

Analysis:
The phrase "their impersonal nature" from the introduction has been re-worded as "insensitive or uncaring."

Topic Sentence 2:
A further problem with emails and texts is that they do not always accurately express the tone which the writer has intended.

Analysis:
The idea of "their inability to capture tone and sarcasm" from the introduction has been paraphrased as "they do not always accurately express the tone which the writer has intended."

Topic Sentence 3:
The danger of having an accessible messaging service readily at hand during times of high emotion is another insidious problem with electronic media.

Analysis:
The phrase "their easy accessibility at times of anger" from the introduction has been expressed here as "readily at hand during times of high emotion."

Also note that the words "further" from topic sentence 2 and "another" from topic sentence 3 improve the flow of the essay by signaling that new ideas are being introduced in these paragraphs.

Writing the Conclusion

Conclusions for CBEST essays can consist of as few as two sentences, provided that the sentences are cohesive, coherent, and well-constructed.

You may want to reiterate certain concepts in the conclusion. However, you should avoid repeating word for word what you have already written.

That is to say that your conclusion can echo your introduction, but you should not repeat the exact phrases you have already used at the start of your essay or in the body paragraphs.

The final sentence of your conclusion can also be used to give advice or to make a prediction about the future. This will give a forward-looking aspect to your essay and will help your writing to end on a strong note.

Writing the Conclusion – Exercise

Look at the underlined words from the introduction to the essay below. Then look at the sample conclusion and identify the words which paraphrase these concepts.
Finally, circle the useful phrases that are used in the sample conclusion.

Introduction:

I agree with the assertion that electronic communication technologies such as email and social media platforms are sometimes used inappropriately. Modern forms of communication such as electronic mail and SMS messaging can cause problems with personal relationships because of three main shortcomings with these media: their <u>impersonal nature</u>, their <u>inability to capture tone and sarcasm</u>, and their <u>easy accessibility at times of anger</u>.

Conclusion:

While email and texts may therefore be useful for certain aspects of our daily lives, these communication methods need to be handled with care in some situations, particularly when they could be seen as insensitive, when it is possible that the recipient might misinterpret the meaning, or when composed at times of personal agitation or stress. The writer of the message should use judgment and common sense in order to avoid the ill feelings that may be caused to the recipient in these cases.

Original wording in the introduction:	Paraphrasing in the conclusion:
impersonal nature	
inability to capture tone and sarcasm	
easy accessibility at times of anger	

Useful phrases in the conclusion:

What advice or prediction is made in the conclusion?

Writing the Conclusion – Answer

Original wording in the introduction:	Paraphrasing in the conclusion:
impersonal nature	could be seen as insensitive
inability to capture tone and sarcasm	the recipient might misinterpret the meaning
easy accessibility at times of anger	composed at times of personal agitation or stress

Useful phrases in the conclusion:

while

particularly

when

in order to

What advice or prediction is made in the conclusion?
The following piece of advice is given in the last sentence of the conclusion to the essay: The writer of the message should use judgment and common sense in order to avoid the ill feelings that may be caused to the recipient in these cases.

SAMPLE EXPOSITORY ESSAY – MODEL ESSAY 1

We reproduce here in full our sample essay, which we have worked on throughout the previous sections of this study guide.

Sample Essay 1:
I agree with the assertion that electronic communication technologies such as email and social media platforms are sometimes used inappropriately. Modern forms of communication such as electronic mail and SMS messaging can cause problems with personal relationships because of three main shortcomings with these media: their impersonal nature, their inability to capture tone and sarcasm, and their easy accessibility at times of anger.

Depending upon the context, the recipient of an email or text message may consider this mode of communication to be insensitive or uncaring. Although email may be practical for conveying straightforward information or facts, electronic messaging would be remarkably inappropriate for events like announcing a death. There is no direct human contact in emails and texts, and during times of loss or tragedy, human warmth and depth of emotion can only truly be conveyed through a phone call, or better still, by talking face to face.

A further problem with emails and texts is that they do not always accurately express the tone which the writer has intended. For instance, it might be possible for the recipient of a sarcastic email message to take its contents literally. The tone of the message may seem abundantly clear to the person who sent it, but sarcastic or ironically humorous utterances can only really be communicated in speech through the tone and inflection of the voice. Without the aid of tone and inflection, certain phrases in an email may come across as demanding, indifferent, or rude.

The danger of having an accessible messaging service readily at hand during times of high emotion is another insidious problem with electronic media. In this day and age, we have heard stories not only of personal break ups that have been conducted by text, but also of employers who fire their staff by email message. Unless the writer of the message has the discipline and self-control to give him or herself a period of reasoned contemplation before sending the communication, he or she might send a regrettable message that can cause irretrievable damage to a relationship.

While email and texts may therefore be useful for certain aspects of our daily lives, these communication methods need to be handled with care in some situations, particularly when they could be seen as insensitive, when it is possible that the recipient might misinterpret the meaning, or when composed at times of personal agitation or stress. The writer of the message should use judgment and common sense in order to avoid the ill feelings that may be caused to the recipient in these cases.

In the next part of the study guide, we include two additional expository essays. You should analyze and study the format and language usage in each expository essay before proceeding to the section on writing the personal experience essay.

SAMPLE EXPOSITORY ESSAY – MODEL ESSAY 2

Thomas Edison once stated: "Restlessness and discontent are the first necessities of progress." Please state whether you agree or disagree with this assertion, giving compelling reasons and examples to support your argument. Your essay should be written for an audience of educated adults.

Like Thomas Edison, I support the view that restlessness and discontent are the first necessities of progress. This simple assertion holds true both on the societal and personal levels.

It is irrefutable that restlessness can lead to progress within society. During the pre-revolutionary period in American history, for example, the settlers in the American colonies became very restless with the way that English law was treating them. This restlessness led to the American Revolutionary War, which witnessed the birth of a myriad of personal and social liberties that American citizens still enjoy today.

In addition, great innovations have come about and continue to come about because of discontent or dissatisfaction with the current state of affairs. Because of this basic principle, many great inventions have been created and many discoveries have occurred. In the nineteenth century, for instance, Louis Braille had an accident at three years of age which caused him to become blind. When he became older, Braille realized that the vast intellectual world of thought and ideas would be closed to him forever unless he devised a system whereby the blind could read. It was this dissatisfaction that led Braille to create the system of type that sight-impaired people around the world utilize today.

Likewise, restlessness and discontent on an individual level can also lead to personal progress. I myself have had life experiences that illustrate this principle. Having worked in an office for many years as an accountant, I realized that I was restless with and discontent in this line of work. This dissatisfaction led me to a journey of self-discovery, culminating in my decision to return to college as a mature student in order to study education.

That is not to say that satisfaction and contentment, whether on a personal or social level, are not to be sought-after. But while satisfaction and contentment can be admirable characteristics in certain ways, these states of mind rarely lead to the social or personal struggles that are necessary in order for change or innovation to occur.

ANALYSIS OF MODEL ESSAY 2:

1. Clear central idea – the writer expresses his central idea clearly in the opening paragraph of the essay when he states: "I support the view that restlessness and discontent are the first necessities of progress [. . . which] holds true both on the societal and personal levels."

2. Well-supported – the essay is well-supported with meaningful explanations and examples. The writer gives examples from Louise Braille's life and the American Revolutionary War, as well as a description of a personal experience to illustrate his viewpoint.

3. Logical organization – the essay is structured very well and is easy to follow and read. The arguments in support of the social implications of the assertion are stated in paragraphs 2 and 3, while the personal implications of the assertion are stated in paragraph 4.

4. Writing conventions – the student has utilized correct grammar, spelling, and punctuation. The essay is also organized into meaningful paragraphs, each of which centers around a key example or point.

5. Syntactic complexity – the structure of the sentences in this essay is very complex. The writer uses linking words and phrases such as when, in addition, during, and likewise to achieve this effect. Please study the underlined words and phrases in the essay and pay attention to how they are being used. If you are unsure of how to construct sentences in this way, please refer to Part 5 of this study guide.

6. Appropriate tone and style – the writer expresses his points of view in a formal, academic way. The level of vocabulary he uses demonstrates that he is writing to an audience of educated adults.

SAMPLE EXPOSITORY ESSAY – MODEL ESSAY 3

> "Money can't buy happiness." Please comment on this statement, giving compelling reasons and examples. Your essay should be written for an audience of educated adults.

I believe in the general concept that money can't buy happiness. Yet, I would also like to add that it is an equally valid point that having a sufficient amount of money can save a person from experiencing a great deal of agony.

Philosophically speaking, happiness and contentment are inner states of mind that are not connected to the tangible world of material existence. Buddhist monks take vows of poverty, and yet they are quite possibly the most content human beings that one could ever encounter. <u>Further</u>, there are known cases of people who have endured unspeakable hardships and <u>nevertheless</u> have continued to have faith in humankind. The *Diary of Anne Frank* is <u>an elucidating example</u> <u>in this case</u>. <u>When</u> reading the writings of this young Jewish girl, one can plainly see that Anne managed to maintain a cheerful outlook throughout her ordeal, <u>in spite of</u> the persecution and hardships she experienced daily while hiding from Nazi soldiers during the Second World War.

<u>Conversely</u>, money itself can be the cause of profound sadness for many people. Nearly every week one can read stories in the media about family feuds over inheritances or lottery winners who find that their friends have turned against them after failing to be given sizeable monetary gifts. <u>What is more</u>, some wealthy people are too preoccupied with increasing their fortunes to stop and enjoy the fruits of their labors by taking a well-deserved vacation or simply spending time relaxing with their families.

It is worth mentioning though, that <u>while</u> money can't buy happiness *per se*, it can alleviate many sources of suffering. Consider, <u>for instance</u>, the person who does not have adequate medical care because he or she cannot afford it. <u>In this case</u>, happiness could be achieved by having ample financial resources to return the person to full health. <u>Of course</u>, even more compelling is the example of children around the globe who must endure the effects of poverty on a daily basis.

<u>Indeed</u>, it seems an existential luxury to engage in philosophical musings over the elusive state of "happiness" when people in many countries around the world will wake up today without the sufficient food, medicine, and drinking water needed for mere survival.

ANALYSIS OF MODEL ESSAY 3:

1. Clear central idea – the writer expresses her central idea clearly in the opening paragraph of the essay when she states: "I believe in the general concept that money can't buy happiness. Yet, I would also like to add that it is an equally valid point that having a sufficient amount of money can save a person from experiencing a great deal of agony." From this sentence, the reader can see that the writer primarily agrees with the essay theme, although she will go on to express certain reservations.

2. Well-supported – the essay is well-supported with meaningful explanations and examples. The writer cites the examples of Buddhist monks, Anne Frank, the discontent of the wealthy, and world poverty to support her position.

3. Logical organization – the essay is structured very well and is easy to follow and read. The arguments in support of the primary assertion are stated in paragraphs 2 and 3, while the writer's reservations are stated in paragraphs 4 and 5.

4. Writing conventions – the student has utilized correct grammar, spelling, and punctuation. The essay is also organized into meaningful paragraphs, each of which focuses on a key point.

5. Syntactic complexity – the structure of the sentences in this essay is very complex. The writer uses linking words such as however, nevertheless, for example, while, and however to achieve this effect. You should again study and analyze the underlined words and phrases in this essay.

6. Appropriate tone and style – the writer expresses her opinion in a clear and assertive way, but she also shows that she can contemplate other viewpoints by using the phrases: it is worth mentioning, I would also like to add that, and it seems.

PART 3 – WRITING THE PERSONAL EXPERIENCE ESSAY

CBEST Personal Experience Essay Structure

As stated at the beginning of this study guide, the purpose of the CBEST personal experience essay is to allow you to express yourself in your writing.

However, that is not to say that the personal experience essay should lack structure or that it can be poorly written.

Even though the structure of the personal experience essay is much freer than that of the expository essay, you may want to follow the organization pattern recommended below.

Paragraph 1

In the first paragraph of the personal experience essay, you should describe and make clear which life event you are going to talk about.

For example, if you are going to talk about a car accident that you were involved in, you should mention that you had an accident.

In other words, you should avoid making a vague statement like "My life was affected by an unfortunate event in my childhood."

The scorer will appreciate clarity as he or she gets focused on your topic at the beginning of your piece.

The body paragraphs

There is no set number of paragraphs that you should use in the personal experience piece.

You could have as few as two and as many as five body paragraphs.

The important point is to be sure that you have expressed yourself clearly and that you have recounted your personal experience more or less completely.

You may find it useful to write about your story chronologically in order to keep your main body paragraphs organized.

The final paragraph

In the last paragraph of your personal response piece, you should bring your personal story to a close.

You can do this by reiterating the reason why the life event was such a significant one for you.

CBEST Personal Experience Essay Tips

<u>Keep it personal</u>

Remember that task 2 is a personal piece. As such, it can and should use first-person pronouns ("I," "me," "my," and "myself,") since you will be recounting a personal experience.

<u>Tone and style</u>

The tone of the personal experience piece will be less formal that that of the expository piece.

<u>Keep it organized</u>

The useful phrases in the next section of the study guide will help you to organize your personal experience piece and keep it focused.

First of all, you should study the list of words provided on the next page.

You should then analyze how these phrases are used in the sample personal experience essays that follow the list of phrases.

Useful Phrases for the Personal Experience Essay

Actions
I began to
I coped by
I embarked upon
Without realizing it, I (did something).

Decisions
I considered (doing something).
I decided to (do something).
I reacted by (doing something).

Discoveries
I came to the conclusion that
I came to the realization that
I discovered that
I found that
I heard that
I realized that
I was informed that

Feelings
The despair I felt
The happiness I felt
The joy I felt
I was distressed when
I was elated when
Much to my delight,
Much to my despair,

Influences
I grew up with
I was influenced by
I was raised to
It was a decisive factor in
He / She demonstrated to me that
The support / love / care / attention I received was invaluable / priceless.

Outcomes
I certainly would have (done something) if I had (known something).
In the end,
It ultimately led to
This instinct remained with me

Thoughts
I contemplated (doing something).
I thought about
I thought that
I mused upon
My thoughts were

Now have a look at the sample personal experience essays that follow and study the useful phrases in each one.

SAMPLE PERSONAL EXPERIENCE ESSAY – MODEL ESSAY 4

> It is often said that every cloud has a silver lining. Describe a difficult situation that you faced in your personal or professional life and explain how you ultimately worked out the problem to your advantage.

When I received word that my application for college had not been accepted, I thought life as I knew it was going to end. "How could life be so unjust?" I mused, as I saw the upcoming academic year stretch out in front of me like a deserted highway. Little did I know that this delay in my academic path would ultimately lead to something truly wonderful.

If I had realized that a simple administrative error on my part was going to delay my studies, I certainly would have been more careful in submitting the necessary forms. I was so self-assured that I was going to be accepted that I had not even bothered to look for work. Nor had I taken into account where I was going to live. Hence, I embarked upon what could have been a year of self-doubt and recrimination.

However, instead of sinking into a quagmire of depression, I decided to take that year as an opportunity to rethink my options. I began to ask myself some hard questions. Did I really want to study in the degree program I had chosen? How committed was I to the idea of financing my own higher education?

I spent weeks scouring the internet for various degree programs and requested a plethora of course catalogues from institutions of higher learning in other states. I then narrowed down my options to nine or ten different colleges.

Invigorated by a new sense of optimism, I requested financial aid and scholarship information from the colleges I had chosen. In the end, three colleges looked the most promising, so I decided to submit applications for admissions, as well as scholarship applications to those places.

Then came the really tough part: waiting for a response. Much to my delight, I was accepted for study at a university in California. I was informed that I would receive a decision about my scholarship application within two months.

The joy I felt when I found out that I had received a full scholarship more than outstripped the agony I had experienced less than a year earlier. Had I not had that setback, I never would have decided to pursue a degree in education.

ANALYSIS OF MODEL ESSAY 4:

1. Clear central idea – the writer expresses his central idea clearly in the opening sentence of the essay when he writes: "When I received word that my application for college had not been accepted, I thought life as I knew it was going to end." Therefore, the examiner can clearly see from the beginning which personal experience the candidate is going to discuss.

2. Well-supported – the essay is well-supported with meaningful explanations and examples. The writer vividly describes his disappointment, as well as the steps he took to overcome his difficulties.

3. Logical organization – the essay is structured very well and is easy to follow and read. The student achieves excellent organization by describing events in chronological order.

4. Writing conventions – the student has utilized correct grammar, spelling, and punctuation. In particular, please notice how to punctuate dialogue and quotations. Remember that punctuation for dialogue should be included within the quotation marks.

5. Syntactic complexity – the structure of the sentences in this essay is very complex. The writer uses linking words and phrases such as if, hence, and nor. The student also begins sentences with past participle phrases. Please see Part 5 of this study guide for advice on how to write sentences in this way.

6. Appropriate tone and style – the writer recounts his life event in an expressive way. Please study the underlined phrases that he uses, which are from the list of useful phrases for the personal experience essay on the previous pages.

SAMPLE PERSONAL EXPERIENCE ESSAY – MODEL ESSAY 5

> Explain what led to your decision to become a teacher. Please support your explanation with reasons and examples. Your essay should be written for an audience of educated adults.

Like many of my colleagues who are about to embark on a career in teaching, I was positively influenced by a teacher who helped me through some difficult personal struggles. The support and concern that this teacher gave me were a major factor in my own decision to enter the teaching profession.

Having had very prominent buck teeth until undergoing orthodontic work in my late teens, I was dubbed "Bugs Bunny" by my classmates in elementary school. Already an awkward and shy youngster, I found that this appellation, although perhaps meant in only jest, exacerbated my lack of self-confidence. I coped with the problem the best I could have at that age: by retreating into my own world of books and reading.

Fortunately, I had one close friend throughout grades 3, 4, and 5. She shared my affinity with reading, and we often exchanged books with each other during summer vacations. Without even realizing it, I was quickly becoming a very proficient reader at a young age. Yet, while improving my reading skill, this habit did little for my self-confidence.

Upon returning to school at the beginning of the sixth grade, my entire self-concept began to change. When I found out that I would be having Mrs. Shelley as my home room and reading teacher, I was absolutely elated. I had heard so many nice things about her and her classes, and she always had a warm smile and time to talk to everyone she met.

Mrs. Shelley started a reading competition for the class that year. Each student had a path on the wall to chart his or her progress. Construction-paper cut-outs of footprints were placed on each student's path each week to represent the number of books that the student had read. Every week, I was reading two or three books, and soon Mrs. Shelley had to prepare another path for me as my first one had become full. My classmates, seeing that my progress was more rapid than theirs, began to call me names in class. Mrs. Shelley then stepped in and saw to it that the taunts of geek, nerd, and brainer were quickly silenced.

Moreover, Mrs. Shelly often complimented me in private after class about my reading skills. Her kindness and sincerity demonstrated to me at an early age the true essence of being a good teacher. Soon my classmates' views of me just didn't matter anymore. I had found something that was important to me: the desire to help other people the way that Mrs. Shelley had helped me. This impulse remained with me throughout middle school and high school, and it was the decisive factor in my own decision to become a teacher.

ANALYSIS OF SAMPLE MODEL 5:

1. Clear central idea – the writer expresses the main idea in the opening sentence of the essay when she states: "Like many of my colleagues who are about to embark on a career in teaching, I was positively influenced by a teacher who helped me through some difficult personal struggles."

2. Well-supported – the essay is well-supported with expressive and personal examples. The writer discusses how she dealt with her derisive nickname, and how this went on to help her become a good reader and ultimately a teacher herself.

3. Logical organization – the essay is structured very well and is easy to follow and read. Like model essay 4, the writer recounts events in chronological order.

4. Writing conventions – the student has utilized correct grammar, spelling, punctuation, and paragraphing.

5. Syntactic complexity – the structure of the sentences in this essay is very complex. The writer uses subordination and linking words such as like, although, fortunately, and however to achieve this effect.

6. Appropriate tone and style – the writer recounts her experience in an extremely expressive way. She achieves this effect by using words and phrases from the list that we have seen previously.

PART 4 – CBEST ENGLISH GRAMMAR REVIEW

Use of the following grammatical conventions is necessary for an effective essay. Please read the following pages carefully, and then have another look at how these conventions are used in the sample essays provided previously in study guide.

Using Correct Grammar and Punctuation

Mechanical conventions are the rules of grammar and punctuation that are necessary in order to write accurately and correctly.

This section covers some of the basic rules of grammar, punctuation, and sentence construction that are assessed in your CBEST essays.

Avoiding Misplaced Modifiers

> Modifiers are descriptive phrases. The modifier should always be placed directly before or after the noun to which it relates.

Now look at the examples.

CORRECT: Like Montana, Wyoming is not very densely populated.
INCORRECT: Like Montana, there isn't a large population in Wyoming.
The phrase "like Montana" is an adjectival phrase that modifies the noun "Wyoming."

Therefore, "Wyoming" must come directly after the comma.

Here are two more examples:
CORRECT: While waiting at the bus stop, a senior citizen was mugged.
INCORRECT: While waiting at the bus stop, a mugging took place.
The adverbial phrase *"while waiting at the bus stop"* modifies the noun phrase "a senior citizen," so this noun phrase needs to come after the adverbial phrase.

Negative Inversion

> When a sentence begins with a negative phrase [no sooner, not only, never, etc.], the present perfect tense [have + past participle] must be used.

CORRECT: Never in my life have I seen such a beautiful sight.
INCORRECT: Never in my life I have seen such a beautiful sight.
This sentence is an example of the inverted sentence structure.

Note that the auxiliary verb "have" must be placed *in front of* the grammatical subject of the sentence [I].

Past Participle Phrases

> Past participles are verb forms that are similar to the past simple tense in their form. In other words, past participles usually end in -ed [in cases of regular verbs].

For example: the past participle of the verb "fluster" is "flustered."

CORRECT: Flustered, Shirley failed her driving test.

INCORRECT: Flustered Shirley failed her driving test.

Remember to put the past participle phrase immediately before or after the noun it modifies.

Also remember to use commas before and after the past participle phrase.

Past Perfect Tense

The past perfect is often used to express an action which has just recently occurred. It can also be used to show that one action preceded another when a sentence describes two actions.

> When describing two actions, the past perfect is used for the action which happened first. The simple past is used for the subsequent action.

CORRECT: Their adversaries in the southern states, the Confederates, had consolidated and called themselves the Confederate States of America.

INCORRECT: Their adversaries in the southern states, the Confederates, consolidated and had called themselves the Confederate States of America.

In other words, the consolidation occurred first. After this, the states began to call themselves the new name.

> The past perfect is often used with the words "just" and "after," and with the phrase "no sooner . . . than." The auxiliary verb must come before the word "just."

CORRECT: We had just arrived, when she decided to leave.

INCORRECT: We just had arrived, when she decided to leave.

Pronoun-Antecedent Agreement

> Pronouns are words like the following: he, she, it, they, and them.
> An antecedent is a phrase that precedes the pronoun in the sentence.

Pronouns must agree with their antecedents.

Now look at the examples below.

CORRECT: Each student needs to bring his or her identification to the placement test.

INCORRECT: Each student needs to bring their identification to the placement test.

The antecedent "each student" is singular, so the singular pronouns "his" or "her" should follow this antecedent.

Pronoun Usage – Correct Use of *Its* and *It's*

> "Its" is a possessive pronoun, while "it's" is a contraction of "it is".

CORRECT: It's high time you started to study.

INCORRECT: Its high time you started to study.

The sentence could also be stated as follows: It is high time you started to study.

Since the contracted form of "it is" can be used in the alternative sentence above, "it's" is the correct form.

CORRECT: A snake sheds its skin at least once a year.

INCORRECT: A snake sheds it's skin at least once a year.

"Its" is a possessive pronoun referring to the snake, so the apostrophe should not be used.

Pronoun Usage – Correct Use of *Their*, *There* and *They're*

> "Their" is a plural possessive pronoun. "There" is used to describe the location of something. "They're" is a contraction of "they are".

CORRECT: Their house is made of brick and concrete.

INCORRECT: There house is made of brick and concrete.

INCORRECT: They're house is made of brick and concrete.

In this case, "their" is the possessive pronoun explaining to whom the house belongs.

CORRECT: He attended college with his cousins living there in California.

INCORRECT: He attended college with his cousins living their in California.

INCORRECT: He attended college with his cousins living they're in California.

"There" is referring to the state of California in the example above, so it is used to talk about the location.

CORRECT: They're away on vacation at the moment.

INCORRECT: Their away on vacation at the moment.

INCORRECT: There away on vacation at the moment.

The sentence could also be written as follows: They are away on vacation at the moment.

"They're" is a contraction of "they are," so the apostrophe needs to be used.

Pronoun Usage – Avoiding "You" and "Your"

> The pronouns "you" and "your" are informal and should generally be avoided in academic writing when referring to a person in general.

FORMAL: Students should plan in advance if they intend to do well on the project.

INFORMAL: You should plan in advance if you intend to do well on the project.

Pronoun Usage – Demonstrative Pronouns

> Demonstrative pronouns include the following words: this, that, these, those

"This" is used for a singular item that is nearby. "That" is used for singular items that are farther away in time or space.

SINGULAR: This book that I have here is really interesting.

PLURAL: That book on the table over there is really interesting.

"These" is used for plural items that are nearby. "Those" is used for plural items that are farther away in time or space.

SINGULAR: These pictures in my purse were taken on our vacation.

PLURAL: Those pictures on the wall were taken on our vacation.

Avoid using "them" instead of "those":

INCORRECT: Them pictures on the wall were taken on our vacation.

Pronoun Usage – Relative Pronouns

Relative pronouns include the following: which, that, who, whom, whose

> "Which" and "that" are used to describe things, and "who" and "whom" are used to describe people. "Whose" is used for people or things.

WHICH: Last night, I watched a romantic-comedy movie which was really funny.

THAT: Last night, I watched a romantic-comedy movie that was really funny.

WHO: Susan always remains calm under pressure, unlike Tom, who is always so nervous.

"Who" is used because we are describing the person. This is known as the nominative case.

WHOM: To whom should the report be given?

"Whom" is used because the person is receiving an action, which in this case is receiving the report. This is known as the accusative case.

WHOSE: I went out for lunch with Marta, whose parents are from Costa Rica.

WHOSE: I went out for lunch yesterday at that new restaurant, whose name I don't remember.

Please be sure to look at the section entitled "Restrictive and Non-restrictive Modifiers" for information on how to use punctuation with relative pronouns.

Punctuation – Avoiding the Parenthetical

> Parentheticals should normally not be used to add extra information to a sentence.

Commas should be used, where possible. Alternatively, two sentences should be written.

CORRECT: Because of drinking beforehand, the driver of the SUV lost control of the vehicle, which overturned on highway and then rolled down the embankment.

CORRECT: The SUV overturned on the highway and then rolled down the embankment after the driver lost control of the vehicle. The accident happened because he had been drinking beforehand.

INCORRECT: The SUV overturned on the highway and then rolled down the embankment after the driver (because of drinking beforehand) lost control of the vehicle.

Punctuation – Using the Apostrophe for Possessive Forms

> Apostrophe placement depends upon whether a word is singular or plural.

For the singular, the apostrophe should be placed before the letter "s."

SINGULAR: Our team's performance was poor at the game last night.

For the plural form, the apostrophe should be placed after the letter "s."

PLURAL: Both teams' performances were poor at the game last night.

Remember that the apostrophe is used in sentences like those above in order to show possession.

Do not use the apostrophe unnecessarily.

INCORRECT: The date's for the events are June 22 and July 5.

INCORRECT: The dates' for the events are June 22 and July 5.

Punctuation – Using Colons and Semicolons

> Colons (:) should be used when giving a list of items. Semicolons (;) should be used to join independent clauses.

COLON: The shop is offering discounts on the following items: DVDs, books, and magazines.

SEMICOLON: I thought I would live in this city forever; then I lost my job.

Please see the section entitled "Punctuation and Independent Clauses" for more information on joining clauses.

Punctuation – Using Commas with Dates and Locations

> Commas should be used after the date and year in dates. Commas should also be used after towns and states.

DATES: On July 4, 1776, the Declaration of Independence was signed.

LOCATIONS: Located in Seattle, Washington, the Space Needle is a major landmark.

Punctuation – Using Commas for Items in a Series

> When using "and" and "or" for more than two items in a series, be sure to use the comma before the words "and" and "or."

CORRECT: You need to bring a tent, sleeping bag, and flashlight.

INCORRECT: You need to bring a tent, sleeping bag and flashlight.

Notice the use of the comma after the word "bag" and before the word "and" in the series.

CORRECT: Students can call, write a letter, or send an email.

INCORRECT: Students can call, write a letter or send an email.

Notice the use of the comma after the word "letter" and before the word "or" in the series.

Punctuation and Independent Clauses – Avoiding Run-On Sentences

> Run-on sentences are those that use commas to join independent clauses together, instead of correctly using the period.

An independent clause contains a grammatical subject and verb. It therefore can stand alone as its own sentence.

The first word of the independent clause should begin with a capital letter, and the clause should be preceded by a period.

CORRECT: I thought I would live in this city forever. Then I lost my job.
INCORRECT: I thought I would live in this city forever, then I lost my job.
"Then I lost my job" is a complete sentence. It has a grammatical subject (I) and a verb (lost). The independent clause must be preceded by a period, and the first word of the new sentence must begin with a capital letter.

Alternatively, an appropriate conjunction can be used to join the independent clauses:

I thought I would live in this city forever, and then I lost my job.

Punctuation and Quotation Marks

> Punctuation should be enclosed within the final quotation mark when giving dialogue.

INCORRECT: "I can't believe you bought a new car", Sam remarked.

CORRECT: "I can't believe you bought a new car," Sam remarked.

The word *exclaimed* shows that the exclamation point is needed in the following examples.

INCORRECT: "I can't believe you bought a new car"! Sam exclaimed.

CORRECT: "I can't believe you bought a new car!" Sam exclaimed.

Restrictive and Non-restrictive Modifiers

> Restrictive modifiers are clauses or phrases that provide essential information that is needed in order to identify the grammatical subject.

Restrictive modifiers should not be preceded by a comma.

Example: My sister who lives in Indianapolis is a good swimmer.

In this case, the speaker has more than one sister, and she is identifying which sister she is talking about by giving the essential information "who lives in Indianapolis."

On the other hand, a non-restrictive modifier is a clause or phrase that provides extra information about a grammatical subject in a sentence. A non-restrictive modifier must be preceded by a comma.

Non-restrictive modifiers are also known as non-essential modifiers.

Example: My sister, who lives in Indianapolis, is a good swimmer.

In this case, the speaker has only one sister. Therefore, the information about her sister's city of residence is not essential in order to identify which sister she is talking about. So, the words "who lives in Indianapolis" form a non-restrictive modifier.

Sentence Fragments

> A sentence fragment is a group of words that does not express a complete train of thought.

CORRECT: I like Christine because she is so nice.

INCORRECT: I like Christine. Because she is so nice.

In the second example, "because she is so nice" is not a complete thought. This idea needs to be joined with the previous clause in order to be grammatically correct.

Subject-Verb Agreement

> Subjects must agree with verbs in number. Be careful with singulars and plurals.

Subject-verb agreement can be confusing when there are intervening words in a sentence.

CORRECT: The flowers in the pots in the garden grow quickly.

INCORRECT: The flowers in the pots in the garden grows quickly.

The grammatical subject in the above sentence is "flowers," not "garden," so the plural form of the verb (*grow*) needs to be used.

CORRECT: Each person in the groups of students needs to pay attention to the instructions.

INCORRECT: Each person in the groups of students need to pay attention to the instructions.

The grammatical subject in the above sentence is "each person," not "students." "Each" is singular and therefore requires the singular form of the verb (*needs*).

Now try the grammar exercises on the next page.

Grammar and Punctuation – Exercises

Each of the sentences below has problems with grammar and punctuation. Find the errors in the sentences and correct them. You may wish to refer to the advice in the previous section as you do the exercise.

The answers are provided on the page following the exercises.

1) I haven't seen her or her sister. Since they went away to college.

2) People who like to get up early in the morning in order to drink more coffee is likely to become easily tired in the afternoon.

3) Hanging from the knob on the bedroom door, Tom thought the new shirt was his favorite.

4) I ran across the street to speak to her, then she surprised me by saying that she had bought a new car.

5) Its common for a magazine to have better sales if it mentions computers, handhelds or other new technology on it's cover.

6) Each student in the class who will take the series of exams on advanced mathematics need to study in advance.

7) Their are several reasons why there having problems with they're children.

8) You have to work hard to succeed at college, so each and every student need to devote time to their studies.

9) Completed on October 28, 1965 the Gateway Arch in St. Louis Missouri is dedicated to Thomas Jefferson who purchased the Louisiana Territory and made the Westward Expansion Movement possible.

10) Before leaving the building at night, please be sure to check the following – the lights, the locks and them storage lockers on the second floor.

11) Student's motivation levels are usually higher when they need to study for final exams.

12) Your phone call (which I told you not to make) interrupted me during an important meeting.

Grammar and Punctuation – Answers

1) I haven't seen her or her sister since they went away to college.

2) People who like to get up early in the morning in order to drink more coffee are likely to become easily tired in the afternoon.

3) Hanging from the knob on the bedroom door, the new shirt was Tom's favorite.

4) I ran across the street to speak to her. Then she surprised me by saying that she had bought a new car.

5) It's common for a magazine to have better sales if it mentions computers, handhelds, or other new technology on its cover.

6) Each student in the class who will take the series of exams on advanced mathematics needs to study in advance.

7) There are several reasons why they're having problems with their children.

8) Students have to work hard to succeed at college, so each and every student needs to devote time to his or her studies.

9) Completed on October 28, 1965, the Gateway Arch in St. Louis, Missouri, is dedicated to Thomas Jefferson, who purchased the Louisiana Territory and made the Westward Expansion Movement possible.

10) Before leaving the building at night, please be sure to check the following: the lights, the locks, and those storage lockers on the second floor.

11) Students' motivation levels are usually higher when they need to study for final exams.

12) Your phone call, which I told you not to make, interrupted me during an important meeting.

PART 5 – DEVELOPING YOUR SENTENCES

How to Use Phrases, Clauses and Cohesive Devices to Develop Your Sentences

In order to perform well on the CBEST essay writing test, you will need to write essays that have advanced and developed sentence structures.

Sentence linking words, sometimes called cohesive devices, are words and phrases that are used in order to combine short sentences together to create more complex sentence structures.

Sentence linking words and phrases fall into three categories: sentence linkers, phrase linkers, and subordinators.

In order to understand how to use these types of sentence linking words and phrases correctly, you will need to know some basics of English grammar.

The basic grammatical principles for these concepts are explained in this section. Be sure to study the examples carefully before you attempt the exercises in the following section of the study guide.

TYPE 1 – SENTENCE LINKERS:
Sentence linkers are used to link two complete sentences together. A complete sentence is one that has a grammatical subject and a verb.

Sentence linkers are usually placed at the beginning of a sentence and are followed by a comma.

They can also be preceded by a semicolon and followed by a comma when joining two sentences together. When doing so, the first letter of the first word of the second sentence must not be capitalized.

<u>Sentence linker examples:</u>

You need to enjoy your time at college. *However*, you should still study hard.
You need to enjoy your time at college; *however*, you should still study hard.

In the examples above, the grammatical subject of the first sentence is "you" and the verb is "need to enjoy".

In the second sentence, "you" is the grammatical subject and "should study" is the verb.

TYPE 2 – PHRASE LINKERS:

In order to understand the difference between phrase linkers and sentence linkers, you must first be able to distinguish a sentence from a phrase.

A phrase linker must be followed by a phrase, while a sentence linker must be followed by a sentence.

The basic distinction between phrases and sentences is that phrases do not have both grammatical subjects and verbs, while sentences contain grammatical subjects and verbs.

Here are some examples of phrases:

Her beauty and grace

Life's little problems

A lovely summer day in the month of June

Working hard

Being desperate for money

Note that the last two phrases above use the –ing form, known in these instances as the present participle.

Present participle phrases, which are often used to modify nouns or pronouns, are sometimes placed at the beginning of sentences as introductory phrases.

Here are some examples of sentences:

Mary worked all day long.

My sister lives in Seattle.

Wintertime is brutal in Milwaukee.

"Mary," "my sister," and "wintertime" are the grammatical subjects of the above sentences.

Remember that verbs are words that show action or states of being, so "worked," "lives," and "is" are the verbs in the three sentences above.

Look at the examples that follow:

Phrase linker example 1 – no comma:

He received a promotion *because of* his dedication to the job.
"His dedication to the job" is a noun phrase.

Phrase linker example 2 – with comma:

Because of his dedication to the job, he received a promotion.

When the sentence begins with the phrase linker, we classify the sentence as an inverted sentence.

Notice that you will need to place a comma between the two parts of the sentence when it is inverted.

TYPE 3 – SUBORDINATORS:

Subordinators must be followed by an independent clause. Subordinators cannot be followed by a phrase.

The two clauses of a subordinated sentence must be separated by a comma.
The structure of independent clauses is similar to that of sentences because independent clauses contain a grammatical subject and a verb.

Subordinator examples:

Although he worked hard, he failed to make his business profitable.

He failed to make his business profitable, *although* he worked hard.

There are two clauses: "He worked hard" and "he failed to make his business profitable."

The grammatical subjects in each clause are the words "he", while the verbs are "worked" and "failed."

Now look at the sentence linking words and phrases below. Note which ones are sentence linkers, which ones are phrase linkers, and which ones are subordinators.
Then refer to the rules above to remember the grammatical principles for sentence linkers, phrase linkers, and subordinators.

Sentence linkers for giving additional information
further
furthermore
apart from this
what is more
in addition
additionally
in the same way
moreover

Sentence linkers for giving examples
for example
for instance
in this case
in particular
more precisely
namely
in brief
in short

Sentence linkers for stating the obvious
obviously
clearly
naturally
of course
surely
after all

Sentence linkers for giving generalizations
in general
on the whole
as a rule
for the most part
generally speaking
in most cases

Sentence linkers for stating causes and effects
thus
accordingly
hence
therefore
in that case
under those circumstances
as a result
for this reason
as a consequence
consequently
in effect

Sentence linkers for concession or unexpected results
however
nevertheless
meanwhile

Sentence linkers for giving conclusions
finally
to conclude
lastly
in conclusion

Sentence linkers for contrast
on the other hand
on the contrary
alternatively
rather

Sentence linkers for paraphrasing or restating
in other words
that is to say
that is

Sentence linkers for showing similarity
similarly
in the same way
likewise

Phrase linkers for giving additional information
besides
in addition to

Phrase linkers for stating causes and effects
because of
due to
owing to

Phrase linkers for concession or unexpected results
despite
in spite of

Phrase linkers for comparison
compared to
like

Phrase linkers for contrast
in contrast to
instead of
rather than
without

Subordinators
although
as
because
but
due to the fact that
even though
since
so
once
unless
until
when
whereas
while
not only . . . but also

Time words that can be used both as phrase linkers and subordinators
after
before

Special cases
yet –"Yet" can be used as both a subordinator and as a sentence linker.
in order to – "In order to" must be followed by the base form of the verb.
thereby – "Thereby" must be followed by the present participle.

We will look at the present participle and base forms in the following exercises.

Sentence Development Exercises

Look at the pairs of sentences in the exercises below. Make new sentences, using the phrase linkers, sentence linkers, and subordinators provided. In many cases, you will need to create one single sentence from the two sentences provided. You may need to change or delete some of the words in the original sentences.

Exercise 1:

    The temperature was quite high yesterday.

    It really didn't feel that hot outside.

Write new sentences beginning as follows:

    a) In spite of . . .

Hint: You need to change the form of the verb "was" in answer (a).

    b) The temperature . . .

You need to include the word "nevertheless" in answer (b). Be careful with punctuation and capitalization in your answer.

Exercise 2:

    Our star athlete didn't receive a gold medal in the Olympics.

    He had trained for competition for several years in advance.

Write new sentences beginning as follows:

    a) Our star athlete . . . .

Answer (a) should contain the word "although."

    b) Despite . . .

Exercise 3:

    There are acrimonious relationships within our extended family.

    Our immediate family decided to go away on vacation during the holiday season to avoid these conflicts.

Write new sentences beginning as follows:

    a) Because of . . .

    b) Because . . .

    c) Due to the fact that . . .

Exercise 4:

    My best friend had been feeling extremely sick for several days.

    She refused to see the doctor.

Write new sentences beginning as follows:

    a) My best friend . . .

Answer (a) should contain the word "however."

    b) My best friend . . .

Answer (b) should contain the word "but."

Be careful with capitalization and punctuation in your answers.

Exercise 5:

    He generally doesn't like drinking alcohol.

    He will do so on social occasions.

Write new sentences beginning as follows:

    a) While . . .

    b) He generally . . .

Answer (b) should contain the word "yet."

Exercise 6:

    The government's policies failed to stimulate spending and expand economic growth.

    The country slipped further into recession.

Write new sentences beginning as follows:

    a) The government's policies . . .

Answer (a) should contain the word "thus."

    b) The government's policies . . .

Answer (b) should contain the word "so."

Exercise 7:

    Students may attend certain classes without fulfilling a prerequisite.

    Students are advised of the benefit of taking at least one non-required introductory course.

Write new sentences beginning as follows:

    a) Even though . . .

    b) Students may attend . . .

Answer (b) should contain the phrase "apart from this."

Exercise 8:

    There have been advances in technology and medical science.

    Infant mortality rates have declined substantially in recent years.

Write new sentences beginning as follows:

    a) Owing to . . .

    b) Since . . .

Exercise 9:

    It was the most expensive restaurant in town.

    It had rude staff and provided the worst service.

Write new sentences beginning as follows:

    a) It was the most . . .

Answer (a) should contain the word "besides."

    b) In addition to . . .

Exercise 10:

Now try to combine these three sentences:

    The judge did not punish the criminal justly.

    He decided to grant a lenient sentence.

    He did not send out a message to deter potential offenders in the future.

Write new sentences as follows:

    a) Instead of . . . and thereby . . .

    b) Rather than . . . in order to . . .

Before you attempt your answer, look for the cause and effect relationships among the three sentences.

In other words, which event came first?

Which ones were second and third in the chain of events?

Also be careful with punctuation in your answers.

Answers to the Sentence Development Exercises

Exercise 1:

Answer (a)

In spite of the temperature being quite high yesterday, it really didn't feel that hot outside.

The words "in spite of" are a phrase linker, not a sentence linker.

That is to say, "in spite of" needs to be followed by a phrase, not a clause.

The verb "was" needs to be changed to "being" in order to form a present participle phrase.

Present participle phrases are made by using the –ing form of the verb. We will see this construction again in some of the following answers.

Answer (b)

There are two possible answers.

The temperature was quite high yesterday. Nevertheless, it really didn't feel that hot outside.

The temperature was quite high yesterday; nevertheless, it really didn't feel that hot outside.

"Nevertheless" is a sentence linker. As such, it needs to be used to begin a new sentence.

Alternatively, the semicolon can be used to join the original sentences. If the semicolon is used, the first letter of the word following it must not be capitalized.

Exercise 2:

Answer (a)

Our star athlete didn't receive a gold medal in the Olympics, although he had trained for competition for several years in advance.

"Although" is a subordinator, so the two sentences can be combined without any changes.

Answer (b)

Despite having trained for competition for several years in advance, our star athlete didn't receive a gold medal in the Olympics.

"Despite" is a phrase linker. As we have seen in answer (a) of exercise 1 above, phrase linkers need to be followed by phrases, not clauses.

The two parts of the sentence are inverted, and the verb "had" needs to be changed to "having" to make the present participle form.

Exercise 3:

Answer (a)

Because of acrimonious relationships within our extended family, our immediate family decided to go away on vacation during the holiday season to avoid these conflicts.

"Because of" is a phrase linker. As such, the subject and verb (there are) need to be removed from the original sentence in order to form a phrase.

Answer (b)

> Because there are acrimonious relationships within our extended family, our immediate family decided to go away on vacation during the holiday season to avoid these conflicts.

Answer (c)

> Due to the fact that there are acrimonious relationships within our extended family, our immediate family decided to go away on vacation during the holiday season to avoid these conflicts.

"Because" and "due to the fact that" are subordinators, so no changes to the original sentences are required.

The phrase "to avoid these conflicts" can be omitted since this idea is already implied by the words "acrimonious relationships."

Exercise 4:

Answer (a)

There are two possible answers.

> My best friend had been feeling extremely sick for several days. However, she refused to see the doctor.

> My best friend had been feeling extremely sick for several days; however, she refused to see the doctor.

Like "nevertheless" in exercise 1, the word "however" is a sentence linker. Remember that sentence linkers usually need to be used at the beginning of a new sentence.

Alternatively, the semicolon can be used to join the original sentences. If the semicolon is used, "however" must not begin with a capital letter and needs to be followed by a comma.

Answer (b)

> My best friend had been feeling extremely sick for several days, but she refused to see the doctor.

"But" is a subordinator, so the two sentences can be combined without any changes.

Exercise 5:

Answer (a)

> While he generally doesn't like drinking alcohol, he will do so on social occasions.

Like the word "although," the word "while" is a subordinator, so no changes to the original sentences are needed.

Answer (b)

"Yet" can be used as both a subordinator and as a sentence linker, so there are three possible answers in this instance.

When used as a sentence linker, the sentence construction is similar to the sentences containing nevertheless" from exercise 1 and "however" from exercise 4.

Accordingly, the two following sentences are possible answers:

> He doesn't like drinking alcohol. Yet, he will do so on social occasions.

> He doesn't like drinking alcohol; yet, he will do so on social occasions.

A third possible answer is to use "yet" as a subordinator.

> He doesn't like drinking alcohol, yet he will do so on social occasions.

The difference is that the third sentence places slightly less emphasis on the particular occasions in which he will drink than the other two sentences.

Exercise 6:

Answer (a)

"Thus" is a sentence linker, so there are two possible answers.

> The government's policies failed to stimulate spending and expand economic growth. Thus, the country slipped further into recession.

> The government's policies failed to stimulate spending and expand economic growth; thus, the country slipped further into recession.

Answer (b)

> The government's policies failed to stimulate spending and expand economic growth, so the country slipped further into recession.

"So" is a subordinator. The two sentences may therefore be joined without any changes.

Exercise 7:

Answer (a)

There are two possible answers.

> Even though students may attend certain classes without fulfilling a prerequisite, they are advised of the benefit of taking at least one non-required introductory course.

> Even though students are advised of the benefit of taking at least one non-required introductory course, they may attend certain classes without fulfilling a prerequisite.

"Even though" is a subordinator, so no changes are needed. It is advisable to change the word "students" to the pronoun "they" on the second part of the new sentence in order to avoid repetition.

The order or the clauses may be changed in the new sentence since there is no cause and effect relationship between the two original sentences.

Answer (b)

There are two possible answers.

> Students may attend certain classes without fulfilling a prerequisite. Apart from this, they are advised of the benefit of taking at least one non-required introductory course.

> Students may attend certain classes without fulfilling a prerequisite; apart from this, they are advised of the benefit of taking at least one non-required introductory course.

"Apart from this" is a sentence linker, so it needs to be used at the beginning of a separate sentence.

Exercise 8:

Answer (a)

> Owing to advances in technology and medical science, infant mortality rates have declined substantially in recent years.

"Owing to" is a phrase linker that shows cause and effect. In this case the cause is advances in technology and medical science, and the effect or result is the decline in infant mortality rates.

Since "owing to" is a phrase linker, the grammatical subject of the original sentence (there) and the verb (have been) are removed when creating the new sentence.

Answer (b)

> Since there have been advances in technology and medical science, infant mortality rates have declined substantially in recent years.

"Since" is a subordinator, so you can combine the sentences without making any changes.

Remember to use the comma between the two parts of the sentence because the clauses have been inverted.

Exercise 9:

Answer (a)

> It was the most expensive restaurant in town, besides having rude staff and providing the worst service.

"Besides" is a phrase linker, so use the present participle form of both verbs in the second original sentence. Accordingly, "had" becomes "having" and "provide" becomes "providing."

Answer (b)

There are two possible answers.

> In addition to being the most expensive restaurant in town, it had rude staff and provided the worst service.

> In addition to having rude staff and providing the worst service, it was the most expensive restaurant in town.

"In addition to" is a phrase linker, so the present participle forms are used in the phrase containing this word.

The order of the original sentences can be changed since there is no cause and effect relationship between these ideas.

Exercise 10:

Answer (a)

> Instead of punishing the criminal justly and thereby sending out a message to deter potential offenders in the future, the judge decided to grant a lenient sentence.

Answer (b)

> Rather than punishing the criminal justly in order to send out a message to deter potential offenders in the future, the judge decided to grant a lenient sentence.

As you can see, answers (a) and (b) above are somewhat similar in their construction.

"Instead of" and "rather than" need to be used with the present particle form (punishing).

"Thereby" must be followed by the present participle form (sending).

However, "in order to" needs to take the base form of the verb (send).

The base form is the verb before any change has been made to it, like making the –ed or –ing forms. The following are examples of base forms of verbs: eat, sleep, work, play.

PART 6 – ESSAY CORRECTION EXERCISES

ESSAY CORRECTION EXERCISE 1

The draft essays that follow contain errors. Choose the correct version of each part of each sentence from the answer choices provided. If the part of the sentence is correct as written, you should choose answer A. The answers are provided at the end of the exercises.

[1] Antarctica is a mysterious and remote continent [2] one which is often forgotten by virtue of its geographical location. [3] Now that the Antarctic is remote and desolate. [4] Nevertheless, an understanding of the organisms that inhabit this continent was critical [5] to our comprehension of the world as a global community. [6] For this reason, the southernmost continent has the source of a great deal of scientific investigation.

[7] Many notable recent research has come from America and Great Britain. [8] The British Antarctic Survey, sponsored with the Natural Environment Research Council of the United Kingdom, [9] and the United States Antarctic Resource Center, a collaborate of the United States Geological Survey Mapping Division and the National Science Foundation, [10] are forerunners in the burgeoning currently field of research in this area.

[11] This corpus of research has resulted in an abundance of factual data on the Antarctic. [12] For example, one now know that more than ninety nine percent of the land is completely covered by snow and ice, [13] which making Antarctica the coldest continent on the planet. [14] This inhospitable climate, has not surprisingly, brought about the adaptation [15] of a plethora of plants and biological organisms on the continent present. [16] An investigation for the sedimentary geological formations provides testimony to the process of adaptation. [17] Ancient sediment's recovered from the bottom of Antarctic lakes, [18] bacteria as well as discovered in ice, [19] has reveal the history of climate change over the past 10,000 years.

Item 1.
- A. Antarctica is a mysterious and remote continent
- B. Antarctica is a mysterious and resounding continent
- C. Antarctica is a mysterious and respectable continent
- D. Antarctica is a mysterious and resistant continent
- E. Antarctica is a mysterious and restrained continent

Item 2.
- A. one which is often forgotten by virtue of its geographical location.
- B. one whose often forgotten by virtue of its geographical location.
- C. that is often forgotten by virtue of its geographical location.
- D. this is often forgotten by virtue of its geographical location.
- E. those are often forgotten by virtue of its geographical location.

Item 3.
- A. Now that the Antarctic is remote and desolate.
- B. Always, the Antarctic is remote and desolate.
- C. Since the Antarctic is remote and desolate.
- D. Indeed, the Antarctic is remote and desolate.
- E. On the other hand, the Antarctic is remote and desolate.

Item 4.
- A. Nevertheless, an understanding of the organisms that inhabit this continent was critical
- B. Nevertheless, an understanding of the organisms that inhabit this continent were critical
- C. Nevertheless, an understanding of the organisms that inhabit this continent is critical

D. Nevertheless, an understanding of the organisms that inhabit this continent are critical
E. Nevertheless, an understanding of the organisms that inhabit this continent are being critical

Item 5.
A. to our comprehension of the world as a global community.
B. to our comprehension at the world as a global community.
C. to our comprehension in the world as a global community.
D. to our comprehension about the world as a global community.
E. to our comprehension for the world as a global community.

Item 6.
A. For this reason, the southernmost continent has the source of a great deal of scientific investigation.
B. For this reason, the southernmost continent has been the source of a great deal of scientific investigation.
C. For this reason, the southernmost continent was the source of a great deal of scientific investigation.
D. For this reason, the southernmost continent has to be the source of a great deal of scientific investigation.
E. For this reason, the southernmost continent had the source of a great deal of scientific investigation.

Item 7.
A. Many notable recent research has come from America and Great Britain.
B. Much notable recent research has come from America and Great Britain.
C. More notable recent research has come from America and Great Britain.
D. More than notable recent research has come from America and Great Britain.
E. As much as notable recent research has come from America and Great Britain.

Item 8.
A. The British Antarctic Survey, sponsored with the Natural Environment Research Council of the United Kingdom,
B. The British Antarctic Survey, sponsored by the Natural Environment Research Council of the United Kingdom,
C. The British Antarctic Survey, sponsored against the Natural Environment Research Council of the United Kingdom,
D. The British Antarctic Survey, sponsored from the Natural Environment Research Council of the United Kingdom,
E. The British Antarctic Survey, sponsored upon the Natural Environment Research Council of the United Kingdom,

Item 9.
A. and the United States Antarctic Resource Center, a collaborate of the United States Geological Survey Mapping Division and the National Science Foundation,
B. And the United States Antarctic Resource Center, a collaborative of the United States Geological Survey Mapping Division and the National Science Foundation,
C. and the United States Antarctic Resource Center, a collaboratively of the United States Geological Survey Mapping Division and the National Science Foundation,
D. and the United States Antarctic Resource Center, a collaboration of the United States Geological Survey Mapping Division and the National Science Foundation,
E. and the United States Antarctic Resource Center, a collaborator of the United States Geological Survey Mapping Division and the National Science Foundation,

Item 10.
- A. are forerunners in the burgeoning currently field of research in this area.
- B. are forerunners in the burgeoning field of currently research in this area.
- C. are currently forerunners in the burgeoning field of research in this area.
- D. are forerunners in the burgeoning field of research in currently this area.
- E. are forerunners in the burgeoning field of research in this currently area.

Item 11.
- A. This corpus of research has resulted in an abundance of factual data on the Antarctic.
- B. This corpus of research was resulted in an abundance of factual data on the Antarctic.
- C. This corpus of research has been resulted in an abundance of factual data on the Antarctic.
- D. This corpus of research was resulting in an abundance of factual data on the Antarctic.
- E. This corpus of research resulting in an abundance of factual data on the Antarctic.

Item 12.
- A. For example, one now know that more than ninety nine percent of the land is completely covered by snow and ice,
- B. For example, we now know that more than ninety nine percent of the land is completely covered by snow and ice,
- C. For example, they now knows that more than ninety nine percent of the land is completely covered by snow and ice,
- D. For example, the community now know that more than ninety nine percent of the land is completely covered by snow and ice,
- E. For example, the research now know that more than ninety nine percent of the land is completely covered by snow and ice,

Item 13.
- A. which making Antarctica the coldest continent on the planet.
- B. which is making Antarctica the coldest continent on the planet.
- C. making Antarctica the coldest continent on the planet.
- D. has made Antarctica the coldest continent on the planet.
- E. that made Antarctica the coldest continent on the planet.

Item 14.
- A. This inhospitable climate, has not surprisingly, brought about the adaptation
- B. This inhospitable climate has, not surprisingly, brought about the adaptation
- C. This inhospitable climate has, not surprisingly; brought about the adaptation
- D. This inhospitable climate has not surprisingly: brought about the adaptation
- E. This inhospitable climate has not surprisingly, brought about the adaptation

Item 15.
- A. of a plethora of plants and biological organisms on the continent present.
- B. of a plethora of plants and biological organisms present on the continent.
- C. of a plethora on the continent of plants and biological organisms present.
- D. of a plethora of plants on the continent and biological organisms present.
- E. of a plethora of plants and on the continent biological organisms present.

Item 16.
- A. An investigation for the sedimentary geological formations provides testimony to the process of adaptation.
- B. An investigation within the sedimentary geological formations provides testimony to the process of adaptation.

C. An investigation at the sedimentary geological formations provides testimony to the process of adaptation.
D. An investigation about the sedimentary geological formations provides testimony to the process of adaptation.
E. An investigation into the sedimentary geological formations provides testimony to the process of adaptation.

Item 17.
A. Ancient sediment's recovered from the bottom of Antarctic lakes,
B. Ancient sediments' recovered from the bottom of Antarctic lakes,
C. Ancient sediments recovered from the bottom of Antarctic lakes,
D. Ancient's sediment recovered from the bottom of Antarctic lakes,
E. Ancient's sediments recovered from the bottom of Antarctic lakes,

Item 18.
A. bacteria as well as discovered in ice,
B. as well as bacteria discovered in ice,
C. bacteria discovered as well as in ice,
D. bacteria discovered in as well as ice,
E. bacteria discovered in ice as well,

Item 19.
A. has reveal the history of climate change over the past 10,000 years.
B. has revealed the history of climate change over the past 10,000 years.
C. have reveal the history of climate change over the past 10,000 years.
D. have revealed the history of climate change over the past 10,000 years.
E. have been revealed the history of climate change over the past 10,000 years.

Item 20.
If the student were to add a paragraph at the end of the essay explaining that the reliability of the research on Antarctica has been disputed, the essay would lose:
A. its academic tone.
B. its clarity and focus.
C. the sense that this topic of current interest.
D. its emphasis on the inhospitality of the Antarctic climate.
E. the sense of importance it places on the scientific evidence.

ESSAY CORRECTION EXERCISE 2

[1] The major significant characteristic of any population is its age-sex structure, [2] defining as the proportion of people of each gender in each different age group. [3] The age-sex structure determines the potential for reproduction, [4] and for example population growth, [5] based on the balance of males and females of child-bearing age inside a population. [6] Thus, the age-sex structure was social policy implications.

[7] For instance, a population with a high proportion of citizens elderly [8] needs to consider its governmental-funded pension schemes and health care systems carefully. [9] As follows: a demographic with a greater percentage of young children should ensure [10] which its educational funding and child welfare policies are implemented efficaciously. [11] Accordingly, as the composition of a population changes against time, [12] the government may need to restate its funding priorities.

[13] For it is possible that a population may have low birth rates [14] resulting an imbalance in the age-sex structure. [15] Low birth rate's might also be attributable to governmental policy that attempts to control the population. [16] Policies are one example of that restrict the number of children a family can have this outcome.

[17] Other possible reason for these types of demographic changes might be unnaturally high death rates, [18] such like in the case of a disease epidemic or natural disaster. [19] Finally, migration is another factor [20] in demographic attrition, because in any population, a certain amount of people, may decide to emigrate, or move to a different country.

Item 1.
- A. The major significant characteristic of any population is its age-sex structure,
- B. The majorly significant characteristic of any population is its age-sex structure,
- C. The most significantly characteristic of any population is its age-sex structure,
- D. The most significant characteristic of any population is its age-sex structure,
- E. The more significant characteristic of any population is its age-sex structure,

Item 2.
- A. defining as the proportion of people of each gender in each different age group.
- B. defined as the proportion of people of each gender in each different age group.
- C. which defining as the proportion of people of each gender in each different age group.
- D. which defined as the proportion of people of each gender in each different age group.
- E. as defined as the proportion of people of each gender in each different age group.

Item 3.
- A. The age-sex structure determines the potential for reproduction,
- B. The age-sex structure determined the potential for reproduction,
- C. The age-sex structure has determined the potential for reproduction,
- D. The age-sex structure had determined the potential for reproduction,
- E. The age-sex structure was determined the potential for reproduction,

Item 4.
- A. and for example population growth,
- B. and so that population growth,
- C. and with regard to population growth,
- D. and it follows that population growth,
- E. and as a consequence population growth,

Item 5.
- A. based on the balance of males and females of child-bearing age inside a population.
- B. based on the balance of males and females of child-bearing age within a population.
- C. based on the balance of males and females of child-bearing age containing a population.
- D. based on the balance of males and females of child-bearing age consisting a population.
- E. based on the balance of males and females of child-bearing age attributing a population.

Item 6.
- A. Thus, the age-sex structure was social policy implications.
- B. Thus, the age-sex structure is social policy implications.
- C. Thus, the age-sex structure has social policy implications.
- D. Thus, the age-sex structure had social policy implications.
- E. Thus, the age-sex structure does social policy implications.

Item 7.
- A. For instance, a population with a high proportion of citizens elderly
- B. For instance, a population with an elderly high proportion of citizens
- C. For instance, a population with a high proportion elderly of citizens
- D. For instance, a population with a high proportion of elderly citizens
- E. For instance, a population with a high elderly proportion of citizens

Item 8.
- A. needs to consider its governmental-funded pension schemes and health care systems carefully.
- B. needs to consider its governmentally-funded pension schemes and health care systems carefully.
- C. needs to consider its funded-governmental pension schemes and health care systems carefully.
- D. needs to consider its funded-governmentally pension schemes and health care systems carefully.
- E. needs to consider its funded governmentally-pension schemes and health care systems carefully.

Item 9.
- A. As follows: a demographic with a greater percentage of young children should ensure
- B. Just as a demographic with a greater percentage of young children should ensure
- C. Conversely, a demographic with a greater percentage of young children should ensure
- D. Despite, a demographic with a greater percentage of young children should ensure
- E. Unless a demographic with a greater percentage of young children should ensure

Item 10.
- A. which its educational funding and child welfare policies are implemented efficaciously.
- B. that its educational funding and child welfare policies are implemented efficaciously.
- C. which it's educational funding and child welfare policies are implemented efficaciously.
- D. that it's educational funding and child welfare policies are implemented efficaciously.
- E. hence its educational funding and child welfare policies are implemented efficaciously.

Item 11.
- A. Accordingly, as the composition of a population changes against time,
- B. Accordingly, as the composition of a population changes for time,
- C. Accordingly, as the composition of a population changes over time,
- D. Accordingly, as the composition of a population changes past time,
- E. Accordingly, as the composition of a population changes as time,

Item 12.
- A. the government may need to restate its funding priorities.
- B. the government may need to re-evaluate its funding priorities.
- C. the government may need to recuperate its funding priorities.
- D. the government may need to cooperate its funding priorities.
- E. the government may need to instigate its funding priorities.

Item 13.
- A. For it is possible that a population may have low birth rates
- B. For this possible that a population may have low birth rates
- C. This is possible that a population may have low birth rates
- D. It is possible that a population may have low birth rates
- E. That is possible that a population may have low birth rates

Item 14.
- A. resulting an imbalance in the age-sex structure.
- B. because an imbalance in the age-sex structure.
- C. due to an imbalance in the age-sex structure.
- D. since an imbalance in the age-sex structure.
- E. in order to imbalance in the age-sex structure.

Item 15.
- A. Low birth rate's might also be attributable to governmental policy that attempts to control the population.
- B. Low birth's rates might also be attributable to governmental policy that attempts to control the population.
- C. Low births' rates might also be attributable to governmental policy that attempts to control the population.
- D. Low birth rates' might also be attributable to governmental policy that attempts to control the population.
- E. Low birth rates might also be attributable to governmental policy that attempts to control the population.

Item 16.
- A. Policies are one example of that restrict the number of children a family can have this outcome.
- B. Policies that restrict are one example of the number of children a family can have this outcome.
- C. Policies that restrict the number of children a family can have this outcome are one example.
- D. Policies that restrict the number of children a family can have are one example of this outcome.
- E. Policies that restrict the number of children a family are one example of this outcome can have.

Item 17.
- A. Other possible reason for these types of demographic changes might be unnaturally high death rates,
- B. Others possible reason for these types of demographic changes might be unnaturally high death rates,
- C. Another possible reason for these types of demographic changes might be unnaturally high death rates,
- D. Anothers possible reason for these types of demographic changes might be unnaturally high death rates,
- E. Another possible reasons for these types of demographic changes might be unnaturally high death rates,

Item 18.
- A. such like in the case of a disease epidemic or natural disaster.
- B. such as in the case of a disease epidemic or natural disaster.
- C. as such as in the case of a disease epidemic or natural disaster.
- D. as its in the case of a disease epidemic or natural disaster.
- E. as much like as in the case of a disease epidemic or natural disaster.

Item 19.
- A. Finally, migration is another factor
- B. Final migration is another factor
- C. Final, migration is another factor
- D. To end, migration is another factor
- E. Conclusively, migration is another factor

Item 20.
- A. in demographic attrition, because in any population, a certain amount of people, may decide to emigrate, or move to a different country.
- B. in demographic attrition because in any population a certain amount of people may decide to emigrate, or move to a different country.
- C. in demographic attrition because, in any population, a certain amount of people may decide to emigrate or move to a different country.

D. in demographic attrition because in any population, a certain amount of people may decide to emigrate, or move to a different country.
E. in demographic attrition because in any population a certain amount of people may decide to emigrate or move to a different country.

Item 21.
Suppose that the student was asked to write an essay, the purpose of which was to explain how the government could rectify current deficiencies in the age-sex structure. Has the student achieved this purpose?
A. Yes, because the student talks about the government's reassessment of funding priorities.
B. Yes, because the student describes the social policy implication of the age-sex structure.
C. Yes, because the student explains the effect of governmental policy on low birth rates.
D. No, because the student fails to provide sufficient examples of the how governmental policy needs to adapt to population changes over time.
E. No, because the student does not enumerate specific solutions that the government could attempt.

ESSAY CORRECTION EXERCISE 3

[1] A group of English separatists known as the Pilgrims first left England to live in Amsterdam, in 1608. [2] After spending a few years in their new city, apart from this, many members of the group [3] felt whose they did not have enough independence. [4] Hence, in 1617, the Pilgrims decided to leave Amsterdam immigrating to America.

[5] More of these separatists were poor farmers [6] whom did not have much education or social status, and, not surprisingly, [7] the group had many financial problems that prevented them for beginning their journey. [8] Thereby their inability to finance themselves caused many disputes and disagreements, [9] the Pilgrims finally managing to obtain financing [10] from a well-known and considerable London businessman named Thomas Weston.

[11] Having secured Weston's monetary support, the group returned to England to pick up some additional passengers, [12] and it boarded a large ship called the Mayflower on September 16, 1620. [13] After 65 days at sea, the pilgrim's reached America.

[14] Plymouth a town about 35 miles southeast of Boston in the New England state of Massachusetts [15] was established by the Pilgrims in December 21, 1620. [16] Even though the early days of this new lives were filled with hope and promise, [17] the harsh winter proved being too much for some of the settlers. [18] Near half of the Pilgrims died during that first winter, [19] but those who lived go on to work hard and prosper.

Item 1.
A. A group of English separatists known as the Pilgrims first left England to live in Amsterdam, in 1608.
B. A group of English separatists known as the Pilgrims first left England to live, in Amsterdam, in 1608.
C. A group of English separatists known as the Pilgrims first left England to live in Amsterdam in 1608.
D. A group of English separatists known as the Pilgrims, first left England to live in Amsterdam, in 1608.
E. A group of English separatists known as the Pilgrims, first left England to live, in Amsterdam in 1608.

Item 2.
A. After spending a few years in their new city, apart from this, many members of the group
B. After spending a few years in their new city, in this case, many members of the group
C. After spending a few years in their new city, namely, many members of the group

D. After spending a few years in their new city, however, many members of the group
E. After spending a few years in their new city, otherwise, many members of the group

Item 3.
A. felt whose they did not have enough independence.
B. felt whom they did not have enough independence.
C. felt which they did not have enough independence.
D. felt that they did not have enough independence.
E. felt in that they did not have enough independence.

Item 4.
A. Hence, in 1617, the Pilgrims decided to leave Amsterdam immigrating to America.
B. Hence, in 1617, the Pilgrims decided to leave Amsterdam to immigrate to America.
C. Hence, in 1617, the Pilgrims decided to leave Amsterdam emigrating to America.
D. Hence, in 1617, the Pilgrims decided to leave Amsterdam to emigrate to America.
E. Hence, in 1617, the Pilgrims decided to leave Amsterdam for migrating to America.

Item 5.
A. More of these separatists were poor farmers
B. Much of these separatists were poor farmers
C. Many of these separatists were poor farmers
D. Many more of these separatists were poor farmers
E. The most of these separatists were poor farmers

Item 6.
A. whom did not have much education or social status, and, not surprisingly,
B. of whom did not have much education or social status, and, not surprisingly,
C. whose did not have much education or social status, and, not surprisingly,
D. which did not have much education or social status, and, not surprisingly,
E. who did not have much education or social status, and, not surprisingly,

Item 7.
A. the group had many financial problems that prevented them for beginning their journey.
B. the group had many financial problems that prevented them to beginning their journey.
C. the group had many financial problems that prevented them from beginning their journey.
D. the group had many financial problems that prevented them against beginning their journey.
E. the group had many financial problems that prevented them with beginning their journey.

Item 8.
A. Thereby their inability to finance themselves caused many disputes and disagreements,
B. Although their inability to finance themselves caused many disputes and disagreements,
C. Nevertheless, their inability to finance themselves caused many disputes and disagreements,
D. Despite their inability to finance themselves caused many disputes and disagreements,
E. In spite of their inability to finance themselves caused many disputes and disagreements,

Item 9.
A. the Pilgrims finally managing to obtain financing
B. the Pilgrims finally managed obtaining financing
C. the Pilgrims finally were managed obtaining financing
D. the Pilgrims finally were managed to obtain financing
E. the Pilgrims finally managed to obtain financing

Item 10.
- A. from a well-known and considerable London businessman named Thomas Weston.
- B. From a well-known and affluent London businessman named Thomas Weston.
- C. from a well-known and unfortunate London businessman named Thomas Weston.
- D. from a well-known and adamant London businessman named Thomas Weston.
- E. from a well-known and insistent London businessman named Thomas Weston.

Item 11.
- A. Having secured Weston's monetary support, the group returned to England to pick up some additional passengers,
- B. To have secured Weston's monetary support, the group returned to England to pick up some additional passengers,
- C. They have secured Weston's monetary support, the group returned to England to pick up some additional passengers,
- D. When they have secured Weston's monetary support, the group returned to England to pick up some additional passengers,
- E. Having secured Weston's monetary support, the group returned to England to pick up some additional passengers

Item 12.
- A. and it boarded a large ship called the Mayflower on September 16, 1620.
- B. and he or she boarded a large ship called the Mayflower on September 16, 1620.
- C. and one boarded a large ship called the Mayflower on September 16, 1620.
- D. and they boarded a large ship called the Mayflower on September 16, 1620.
- E. and those boarded a large ship called the Mayflower on September 16, 1620.

Item 13.
- A. After 65 days at sea, the pilgrim's reached America.
- B. After 65 days at sea, the Pilgrims' reached America.
- C. After 65 days at sea, the Pilgrims reached America.
- D. After 65 days at sea, Pilgrim's reached America.
- E. After 65 days at sea, Pilgrims' reached America.

Item 14.
- A. Plymouth a town about 35 miles southeast of Boston in the New England state of Massachusetts
- B. Plymouth, a town about 35 miles southeast of Boston in the New England state of Massachusetts,
- C. Plymouth, a town about 35 miles southeast of Boston in the New England, state of Massachusetts
- D. Plymouth, a town about 35 miles southeast of Boston in the New England, state of Massachusetts,
- E. Plymouth, a town about 35 miles southeast of Boston, in the New England, state of Massachusetts,

Item 15.
- A. was established by the Pilgrims in December 21, 1620.
- B. was established by the Pilgrims on December 21, 1620.
- C. was established by the Pilgrims at December 21, 1620.
- D. was established by the Pilgrims upon December 21, 1620.
- E. was established by the Pilgrims during December 21, 1620.

Item 16.
- A. Even though the early days of this new lives were filled with hope and promise,
- B. Even though the early days of that new lives were filled with hope and promise,

    C. Even though the early days of their new lives were filled with hope and promise,
    D. Even though the early days of these new live were filled with hope and promise,
    E. Even though the early days of those new live were filled with hope and promise,

Item 17.
    A. the harsh winter proved being too much for some of the settlers.
    B. the harsh winter proved to be too much for some of the settlers.
    C. the harsh winter proved to being too much for some of the settlers.
    D. the harsh winter proved been too much for some of the settlers.
    E. the harsh winter proved to been too much for some of the settlers.

Item 18.
    A. Near half of the Pilgrims died during that first winter,
    B. Nearly half of the Pilgrims died during that first winter,
    C. Nearly of half of the Pilgrims died during that first winter,
    D. Near of half of the Pilgrims died during that first winter,
    E. Almost near half of the Pilgrims died during that first winter,

Item 19.
    A. but those who lived go on to work hard and prosper.
    B. but those who lived goes on to work hard and prosper.
    C. but those who lived going on to work hard and prosper.
    D. but those who lived went on to work hard and prosper.
    E. but those who lived had went on to work hard and prosper.

Item 20.
If the student removed the last sentence of the essay, how would this affect the essay?
    A. The essay would have more emphasis on the hardships of the Pilgrims.
    B. The comments on the early days of the Pilgrims would have increased importance.
    C. The historical account of the Pilgrims would lack continuity.
    D. The essay would lack a sense of focus.
    E. The essay would lack a proper conclusion.

ESSAY CORRECTION EXERCISE 4

[1] In 1929 that electrical activity in the human brain was first discovered. [2] Hans Berger, the German psychiatrist made the discovery, [3] was despondent to find out, in contrast to, that his research was quickly dismissed by many other scientists.

[4] The work of Berger was confirmed three years later, in 1932, when Edgar Adrian a Briton, [5] clearly demonstrated that the brain, like the heart, is profuse in its electrical activity. [6] Because of Adrian's work, it know that the electrical impulses [7] in the brain called brain waves are a mixture of four different frequencies, [8] that are based on the number of electrical impulses [9] that occurring in the brain per second.

[10] Accordingly, there are four types of brain waves as follows, alpha, beta, delta, and theta. [11] Alpha waves occur in a state of relaxation, while beta waves occur when a person is alert. [12] In addition, delta waves take place for sleep, but they can also occur dysfunctionally when the brain has been severely damaged. [13] Finally, theta waves are a frequency of [14] somewhere in between alpha and delta. [15] Seems that the purpose of theta waves is solely to facilitate the combination of the other brain waves.

[16] The whole notion of brain waves feeds into the current controversy about brain death. [17] Some believe that brain death is characterized by the failure of the cerebral cortex to function. [18] On the other hand, anothers say that mere damage to the cerebral cortex is not enough. [19] They assert

that brain stem function must also cease before can a person be declared dead because the cerebral cortex is responsible for other bodily processes.

Item 1.
- A. In 1929 that electrical activity in the human brain was first discovered.
- B. It in 1929 that electrical activity in the human brain was first discovered.
- C. It was in 1929 that electrical activity in the human brain was first discovered.
- D. It in 1929 was that electrical activity in the human brain was first discovered.
- E. That in 1929 electrical activity in the human brain was first discovered.

Item 2.
- A. Hans Berger, the German psychiatrist made the discovery,
- B. Hans Berger, the German psychiatrist had made the discovery,
- C. Hans Berger, the German psychiatrist who made the discovery,
- D. Hans Berger, the German psychiatrist whom made the discovery,
- E. Hans Berger, the German psychiatrist which made the discovery,

Item 3.
- A. was despondent to find out, in contrast to, that his research was quickly dismissed by many other scientists.
- B. was despondent to find out, likewise, that his research was quickly dismissed by many other scientists.
- C. was despondent to find out, but, that his research was quickly dismissed by many other scientists.
- D. was despondent to find out, though, that his research was quickly dismissed by many other scientists.
- E. was despondent to find out, although, that his research was quickly dismissed by many other scientists.

Item 4.
- A. The work of Berger was confirmed three years later, in 1932, when Edgar Adrian a Briton,
- B. The work of Berger was confirmed three years later, in 1932, when Edgar Adrian, a Briton,
- C. The work of Berger was confirmed three years later, in 1932, when Edgar Adrian a Briton
- D. The work of Berger was confirmed three years later, in 1932, when Edgar Adrian a Briton;
- E. The work of Berger was confirmed three years later, in 1932, when Edgar Adrian, a Briton;

Item 5.
- A. clearly demonstrated that the brain, like the heart, is profuse in its electrical activity.
- B. demonstrated that the clearly brain, like the heart, is profuse in its electrical activity.
- C. demonstrated that the brain, like clearly the heart, is profuse in its electrical activity.
- D. demonstrated that the brain, like the heart clearly, is profuse in its electrical activity.
- E. demonstrated that the brain, like the heart, is profuse clearly in its electrical activity.

Item 6.
- A. Because of Adrian's work, it know that the electrical impulses
- B. Because of Adrian's work, it known that the electrical impulses
- C. Because of Adrian's work, it is known that the electrical impulses
- D. Because of Adrian's work, we known that the electrical impulses
- E. Because of Adrian's work, one known that the electrical impulses

Item 7.
- A. in the brain called brain waves are a mixture of four different frequencies,
- B. in the brain, called brain waves are a mixture of four different frequencies,
- C. in the brain called brain waves, are a mixture of four different frequencies,
- D. in the brain, called brain waves, are a mixture of four different frequencies,
- E. in the brain, called brain waves, are a mixture, of four different frequencies,

Item 8.
- A. that are based on the number of electrical impulses
- B. that based on the number of electrical impulses
- C. which are based on the number of electrical impulses
- D. which based on the number of electrical impulses
- E. are based on the number of electrical impulses

Item 9.
- A. that occurring in the brain per second.
- B. that occurred in the brain per second.
- C. that had occurred in the brain per second.
- D. that have occurrence in the brain per second.
- E. that occur in the brain per second.

Item 10.
- A. Accordingly, there are four types of brain waves as follows, alpha, beta, delta, and theta.
- B. Accordingly, there are four types of brain waves as follows: alpha, beta, delta, and theta.
- C. Accordingly, there are four types of brain waves as follows; alpha, beta, delta, and theta.
- D. Accordingly, there are four types of brain waves as follows alpha, beta, delta, and theta.
- E. Accordingly, there are four types of brain waves as follows. Alpha, beta, delta, and theta.

Item 11.
- A. Alpha waves occur in a state of relaxation, while beta waves occur when a person is alert.
- B. Alpha waves occur in a state of relaxation, rather beta waves occur when a person is alert.
- C. Alpha waves occur in a state of relaxation, rather than beta waves occur when a person is alert.
- D. Alpha waves occur in a state of relaxation, instead of waves occur when a person is alert.
- E. Alpha waves occur in a state of relaxation, as for beta waves occur when a person is alert.

Item 12.
- A. In addition, delta waves take place for sleep, but they can also occur dysfunctionally when the brain has been severely damaged.
- B. In addition, delta waves take place during sleep, but they can also occur dysfunctionally when the brain has been severely damaged.
- C. In addition, delta waves take place since sleep, but they can also occur dysfunctionally when the brain has been severely damaged.
- D. In addition, delta waves take place with sleep, but they can also occur dysfunctionally when the brain has been severely damaged.
- E. In addition, delta waves take place at sleep, but they can also occur dysfunctionally when the brain has been severely damaged.

Item 13.
- A. Finally, theta waves are a frequency of
- B. Finally, theta waves are of a frequency
- C. Finally, theta waves of are a frequency
- D. Finally, of theta waves are a frequency
- E. Finally, theta waves are a of frequency

Item 14.
- A. somewhere in between alpha and delta.
- B. somewhere with between alpha and delta.
- C. somewhere in besides alpha and delta.
- D. somewhere at between alpha and delta.
- E. somewhere at besides alpha and delta.

Item 15.
- A. Seems that the purpose of theta waves is solely to facilitate the combination of the other brain waves.
- B. Seemingly that the purpose of theta waves is solely to facilitate the combination of the other brain waves.
- C. It seemingly that the purpose of theta waves is solely to facilitate the combination of the other brain waves.
- D. It is seemingly that the purpose of theta waves is solely to facilitate the combination of the other brain waves.
- E. It seems that the purpose of theta waves is solely to facilitate the combination of the other brain waves.

Item 16.
- A. The whole notion of brain waves feeds into the current controversy about brain death.
- B. The whole notion of brain waves feeds at the current controversy about brain death.
- C. The whole notion of brain waves feeds with the current controversy about brain death.
- D. The whole notion of brain waves feeds against the current controversy about brain death.
- E. The whole notion of brain waves feeds for the current controversy about brain death.

Item 17.
- A. Some believe that brain death is characterized by the failure of the cerebral cortex to function.
- B. Some people's belief that brain death is characterized by the failure of the cerebral cortex to function.
- C. Some peoples' belief that brain death is characterized by the failure of the cerebral cortex to function.
- D. Certain peoples believe that brain death is characterized by the failure of the cerebral cortex to function.
- E. Certain believe that brain death is characterized by the failure of the cerebral cortex to function.

Item 18.
- A. On the other hand, anothers say that mere damage to the cerebral cortex is not enough.
- B. On the other hand, another say that mere damage to the cerebral cortex is not enough.
- C. On the other hand, others say that mere damage to the cerebral cortex is not enough.
- D. On the other hand, other say that mere damage to the cerebral cortex is not enough.
- E. On the other hand, other's say that mere damage to the cerebral cortex is not enough.

Item 19.
    A. They assert that brain stem function must also cease before can a person be declared dead because the cerebral cortex is responsible for other bodily processes.
    B. They assert that brain stem function must also cease before a person can be declared dead because the cerebral cortex is responsible for other bodily processes.
    C. They assert that brain stem function must also cease before may a person be declared dead because the cerebral cortex is responsible for other bodily processes.
    D. They assert that brain stem function must also cease before might a person can be declared dead because the cerebral cortex is responsible for other bodily processes.
    E. They assert that brain stem function must also cease before a person declared dead because the cerebral cortex is responsible for other bodily processes.

Item 20.
Imagine that the student would like to add the following sentence to the essay. What is the best location for this sentence?
Therefore, for these myriad reasons, it has become very important to measure brain activity.
    A. At the end of the first paragraph.
    B. At the end of the second paragraph.
    C. At the end of the third paragraph.
    D. At the beginning of the last paragraph.
    E. At the end of the last paragraph.

ESSAY CORRECTION EXERCISE 5

[1] Cancer, a group of mainly than 100 different types of disease, [2] occurs where cells in the body begin to divide abnormally and continue dividing and forming more cells without control or order. [3] All internal organs of the body consist of cells, which normally divide to produce more cells when the body requires them. [4] This is a natural, orderly process, that keeps human beings healthy.

[5] If a cell divides when is not necessary, a large growth called a tumor can form. [6] These tumors can usually be removed, and in many cases, they do not recurrence. [7] Unfortunately, in some cases the cancer at the original tumor spreads. [8] The spread of cancer in such way is called metastasis.

[9] There are some factors which are being known to increase the risk of cancer. [10] Smoking is the single cause largest of death from cancer in the United States. [11] One-third of the death's from cancer each year are related to smoking, [12] making tobacco use the most preventable cause of death in this country.

[13] Choice of food can also be link to cancer. [14] Research shows that there are a link between high-fat food and certain cancers, and being seriously overweight is also a cancer risk. [15] Cancer risk can be reduced with a cut down on fatty food and eating generous amounts of fruit and vegetables.

Item 1.
    A. Cancer, a group of mainly than 100 different types of disease,
    B. Cancer, a group of more than 100 different types of disease,
    C. Cancer, a group of 100 more different types of disease,
    D. Cancer, a group of mostly than 100 different types of disease,
    E. Cancer, a group of almost than 100 different types of disease,

Item 2.
    A. occurs where cells in the body begin to divide abnormally and continue dividing and forming more cells without control or order.
    B. occurs which cells in the body begin to divide abnormally and continue dividing and forming more cells without control or order.
    C. occurs in which cells in the body begin to divide abnormally and continue dividing and forming more cells without control or order.

D. occurs when cells in the body begin to divide abnormally and continue dividing and forming more cells without control or order.
E. occurs once when cells in the body begin to divide abnormally and continue dividing and forming more cells without control or order.

Item 3.
A. All internal organs of the body consist of cells, which normally divide to produce more cells when the body requires them.
B. All internal organs of the body consist of cells, which divide to normally produce more cells when the body requires them.
C. All internal organs of the body consist of cells, which divide to produce more normally cells when the body requires them.
D. All internal organs of the body consist of cells, which divide to produce more cells when normally the body requires them.
E. All internal organs of the body consist of cells, which divide to produce more cells when the body requires them normally.

Item 4.
A. This is a natural, orderly process, that keeps human beings healthy.
B. This is a natural, orderly process that keeps human beings healthy.
C. This is a natural orderly process, that keeps human beings healthy.
D. This is a natural orderly process that keeps human beings healthy.
E. This is a natural orderly, process that keeps human beings healthy.

Item 5.
A. If a cell divides when is not necessary, a large growth called a tumor can form.
B. If a cell divides when they are not necessary, a large growth called a tumor can form.
C. If a cell divides when it is not necessary, a large growth called a tumor can form.
D. If a cell divides when are not necessary, a large growth called a tumor can form.
E. If a cell divides when that not necessary, a large growth called a tumor can form.

Item 6.
A. These tumors can usually be removed, and in many cases, they do not recurrence.
B. These tumors can usually be removed, and in many cases, they do not make recurrence.
C. These tumors can usually be removed, and in many cases, they do not recurring.
D. These tumors can usually be removed, and in many cases, they do not are recurred.
E. These tumors can usually be removed, and in many cases, they do not recur.

Item 7.
A. Unfortunately, in some cases the cancer at the original tumor spreads.
B. Unfortunately, in some cases the cancer from the original tumor spreads.
C. Unfortunately, in some cases the cancer with the original tumor spreads.
D. Unfortunately, in some cases the cancer for the original tumor spreads.
E. Unfortunately, in some cases the cancer below the original tumor spreads.

Item 8.
A. The spread of cancer in such way is called metastasis.
B. The spread of cancer in such a way is called metastasis.
C. The spread of cancer in such ways is called metastasis.
D. The spread of cancer in such like way is called metastasis.
E. The spread of cancer in such like ways is called metastasis.

Item 9.
- A. There are some factors which are being known to increase the risk of cancer.
- B. There are some factors which are know to increase the risk of cancer.
- C. There are some factors which are knowing to increase the risk of cancer.
- D. There are some factors which are known to increase the risk of cancer.
- E. There are some factors which have known to increase the risk of cancer.

Item.10
- A. Smoking is the single cause largest of death from cancer in the United States.
- B. Smoking is the single cause of largest death from cancer in the United States.
- C. Smoking is the single cause of death largest from cancer in the United States.
- D. Smoking is the single cause of death from cancer largest in the United States.
- E. Smoking is the largest single cause of death from cancer in the United States.

Item 11.
- A. One-third of the death's from cancer each year are related to smoking,
- B. One-third of the deaths' from cancer each year are related to smoking,
- C. One-third of the deaths from cancer each year are related to smoking,
- D. One-third of cancer's deaths each year are related to smoking,
- E. One-third of cancers' deaths each year are related to smoking,

Item 12.
- A. making tobacco use the most preventable cause of death in this country.
- B. which making tobacco use the most preventable cause of death in this country.
- C. made tobacco use the most preventable cause of death in this country.
- D. which will be making tobacco use the most preventable cause of death in this country.
- E. in making tobacco use the most preventable cause of death in this country.

Item 13.
- A. Choice of food can also be link to cancer.
- B. Choice of food can also be linking to cancer.
- C. Choice of food can also be linked to cancer.
- D. Choice of food can also been linked to cancer.
- E. Choice of food can also link to cancer.

Item 14.
- A. Research shows that there are a link between high-fat food and certain cancers, and being seriously overweight is also a cancer risk.
- B. Research shows that there is a link between high-fat food and certain cancers, and being seriously overweight is also a cancer risk.
- C. Research shows that there's links between high-fat food and certain cancers, and being seriously overweight is also a cancer risk.
- D. Research shows that there is existing a link between high-fat food and certain cancers, and being seriously overweight is also a cancer risk.
- E. Research shows that there in existence a link between high-fat food and certain cancers, and being seriously overweight is also a cancer risk.

Item 15.
- A. Cancer risk can be reduced with a cut down on fatty food and eating generous amounts of fruit and vegetables.
- B. Cancer risk can be reduced with cutting down on fatty food and eating generous amounts of fruit and vegetables.
- C. Cancer risk can be reduced with cutting down fatty food and eating generous amounts of fruit and vegetables.

D. Cancer risk can be reduced by cutting down on fatty food and eating generous amounts of fruit and vegetables.
E. Cancer risk can be reduced by cut down on fatty food and eating generous amounts of fruit and vegetables.

Item 16.
Suppose the student wants to include an admonition to the reader about how he or she can prevent cancer risks. Which sentence, if added to the end of the essay, would achieve this purpose?
A. Accordingly, the government needs to act now to help improve the health of the country.
B. Militating against the causes of cancer is a difficult but necessary task.
C. It is therefore the responsibility of each individual to try to mitigate cancer risk by living a healthy lifestyle.
D. However, these deaths could easily have been avoided.
E. Nevertheless, most people agree that trying to prevent cancer risk is extremely important.

ASWERS TO THE ESSAY CORRECTION EXERCISES

Antarctica Essay

1. A
2. C
3. D
4. C
5. A
6. B
7. B
8. B
9. D
10. C
11. A
12. B
13. C
14. B
15. B
16. E
17. C
18. B
19. D
20. E

Population Age-Sex Structure Essay

1. D
2. B
3. A
4. E

5. B
6. C
7. D
8. B
9. C
10. B
11. C
12. B
13. D
14. C
15. E
16. D
17. C
18. B
19. A
20. D
21. E

The Pilgrims Essay

1. C
2. D
3. D
4. D
5. C
6. E
7. C
8. B
9. E

10. B

11. A

12. D

13. C

14. B

15. B

16. C

17. B

18. B

19. D

20. E

Brain Wave Research Essay

1. C

2. C

3. D

4. B

5. A

6. C

7. D

8. C

9. E

10. B

11. A

12. B

13. B

14. A

15. E

16. A

17. A

18. C

19. B

20. E

Cancer Risk Essay

1. B
2. D
3. A
4. B
5. C
6. E
7. B
8. B
9. D
10. E
11. C
12. A
13. C
14. B
15. D
16. C

www.ingramcontent.com/pod-product-compliance
Lightning Source LLC
Chambersburg PA
CBHW081741100526
44592CB00015B/2248